Proceedings

Ein stetig steigender Fundus an Informationen ist heute notwendig, um die immer komplexer werdende Technik heutiger Kraftfahrzeuge zu verstehen. Funktionen, Arbeitsweise, Komponenten und Systeme entwickeln sich rasant. In immer schnelleren Zyklen verbreitet sich aktuelles Wissen gerade aus Konferenzen, Tagungen und Symposien in die Fachwelt. Den raschen Zugriff auf diese Informationen bietet diese Reihe Proceedings, die sich zur Aufgabe gestellt hat, das zum Verständnis topaktueller Technik rund um das Automobil erforderliche spezielle Wissen in der Systematik aus Konferenzen und Tagungen zusammen zu stellen und als Buch in Springer.com wie auch elektronisch in Springer Link und Springer Professional bereit zu stellen. Die Reihe wendet sich an Fahrzeug- und Motoreningenieure sowie Studierende, die aktuelles Fachwissen im Zusammenhang mit Fragestellungen ihres Arbeitsfeldes suchen. Professoren und Dozenten an Universitäten und Hochschulen mit Schwerpunkt Kraftfahrzeug- und Motorentechnik finden hier die Zusammenstellung von Veranstaltungen, die sie selber nicht besuchen konnten. Gutachtern, Forschern und Entwicklungsingenieuren in der Automobil- und Zulieferindustrie sowie Dienstleistern können die Proceedings wertvolle Antworten auf topaktuelle Fragen geben.

Today, a steadily growing store of information is called for in order to understand the increasingly complex technologies used in modern automobiles. Functions, modes of operation, components and systems are rapidly evolving, while at the same time the latest expertise is disseminated directly from conferences, congresses and symposia to the professional world in ever-faster cycles. This series of proceedings offers rapid access to this information, gathering the specific knowledge needed to keep up with cutting-edge advances in automotive technologies, employing the same systematic approach used at conferences and congresses and presenting it in print (available at Springer.com) and electronic (at Springer Link and Springer Professional) formats. The series addresses the needs of automotive engineers, motor design engineers and students looking for the latest expertise in connection with key questions in their field, while professors and instructors working in the areas of automotive and motor design engineering will also find summaries of industry events they weren't able to attend. The proceedings also offer valuable answers to the topical questions that concern assessors, researchers and developmental engineers in the automotive and supplier industry, as well as service providers.

Weitere Bände in der Reihe http://www.springer.com/series/13360

Torsten Bertram
(Hrsg.)

Automatisiertes Fahren 2019

Von der Fahrerassistenz zum autonomen Fahren
5. Internationale ATZ-Fachtagung

Springer Vieweg

Hrsg.
Torsten Bertram
Technische Universität Dortmund
Dortmund, Deutschland

ISSN 2198-7432 ISSN 2198-7440 (electronic)
Proceedings
ISBN 978-3-658-27989-9 ISBN 978-3-658-27990-5 (eBook)
https://doi.org/10.1007/978-3-658-27990-5

Die Deutsche Nationalbibliothek verzeichnet diese Publikation in der Deutschen Nationalbibliografie; detaillierte bibliografische Daten sind im Internet über http://dnb.d-nb.de abrufbar.

Springer Vieweg

Verantwortlich im Verlag: Markus Braun

Springer Vieweg ist ein Imprint der eingetragenen Gesellschaft Springer Fachmedien Wiesbaden GmbH und ist ein Teil von Springer Nature.
Die Anschrift der Gesellschaft ist: Abraham-Lincoln-Str. 46, 65189 Wiesbaden, Germany

Vorwort

Die fünfte ATZ-Fachtagung „Automatisiertes Fahren – Von der Fahrerassistenz zum autonomen Fahren“ 2019 setzt die erfolgreiche Serie der Fachtagungen Fahrerassistenzsysteme seit 2015 fort und zeigt durch die Umbenennung der Fachtagung von „Fahrerassistenzsysteme“ in „Automatisiertes Fahren“ den Technologiefortschritt gemäß der Einteilung der automatisierten Fahrfunktionen nach SAE Level und die Fokussierung der Themen der Fachtagung.

Die Automatisierung der Fahrfunktionen, die Elektrifizierung des Antriebsstrangs, die Vernetzung der Verkehrsteilnehmer und die Individualisierung der Verkehrssysteme charakterisieren die Mobilität der Zukunft. Diese wird sich grundsätzlich von der heute gewohnten Mobilität unterscheiden.

Die angesprochenen Entwicklungen führen zu einer höheren Verkehrssicherheit und einem zunehmenden Komfort, da der Mensch als Risikofaktor bei der Ausübung der Fahrfunktionen beim automatisierten Fahren zunehmend wegfällt und er durch den Wandel vom Fahrer zum Passagier nun die Reisezeit frei nutzen kann. Ein sich damit einstellender gleichmäßigerer Verkehrsfluss führt auch zu einer Steigerung der Verkehrseffizienz und zu einer besseren Umweltbilanz.

Der Wechsel vom manuellen Fahren über das teil-, hoch- und vollautomatisierte Fahren hin zum fahrerlosen Fahrzeug wird sich auf der Autobahn, in der ländlichen Umgebung und in der urbanen Region auf unterschiedlichen Zeitachsen und auch hinsichtlich der beherrschbaren Szenarien verschieden entwickeln, da sich die Komplexität und Kritikalität möglicher Verkehrssituationen deutlich unterscheiden.

Das Themenspektrum der Fachtagung zum automatisierten Fahren reicht 2019 von den erforderlichen Methoden und Kompetenzen bis hin zur Nutzerakzeptanz.

In vier Übersichtsbeiträgen, einer Podiumsdiskussion zum Thema „Reality check for automated driving – how we will drive in 2025?“ und in acht Themengebieten: Methoden und Prozesse, Absicherung, Autonome

Versicherungsvisionen, Haftungsfragen/ Fahrzeugdaten, Künstliche Intelligenz, Umfelderkennung, Daten und Vernetzung sowie Nutzer und Akzeptanz werden viele Aspekte des automatisierten Fahrens beleuchtet und zur Diskussion gestellt.

Die Übersichtsbeiträge geben Einblicke in den Trend beim automatisierten Fahren im Bereich der Nutzfahrzeuge generell und zeigen, wie sich die Mobilität in Smart Cities in China, Japan und Europa aus Sicht eines Fahrzeugherstellers und Zulieferers entwickeln wird. Der Rolle und Bedeutung der künstlichen Intelligenz für das automatisiert fahrende Fahrzeug und die zu schaffenden Voraussetzungen bei deren Einsatz im Fahrzeug wird in einem eigenen Übersichtsbeitrag aus Sicht eines Automobilzulieferers aufgezeigt.

Fahrzeuge mit automatisierten Fahrfunktionen gemäß SAE Level 3 und höher sind in der Lage zu sehen – 360-Grad-Wahrnehmung der Umgebung über Sensoren –, zu denken – vorausschauend und situationsgerecht einen angemessenen Handlungsablauf über Situationsinterpretation, Schlussfolgerung, Planung, Planerkennung, Kommunikation sowie Kollaboration zu generieren – und zu handeln – Handlungen sicher und zuverlässig über Aktoren auszuführen –.

Bei der Selbstregulation (Wahrnehmung und Interpretation, Lernen und Schlussfolgern, Planung und Planerkennung, Kommunikation und Kollaboration) kann die Künstliche Intelligenz einen bedeutungsvollen Beitrag leistet. Mit dem Machine Learning ist dem automatisiert fahrenden Fahrzeug ein Werkzeug gegeben, dass bei einer umfangreichen Anzahl an Messungen aus diesen Daten komplexe Zusammenhänge erkennt. Mit zunehmender Integration der Künstlichen Intelligenz in das automatisiert fahrende Fahrzeug wird dieses mehr und mehr ohne menschliche Steuerung sowie Regelung und durch detaillierte situationsausgerichtete Programme ein vorgegebenes Ziel selbständig und an die aktuelle Situation angepasst erreichen.

Die Beiträge zu den Themengebieten Methoden und Prozesse sowie Absicherung widmen sich der Integration der Verifikation sowie Validierung von komplexen Systemen bereits in frühen Entwicklungsphasen. Daneben stehen die Generierung von Daten der Umweltwahrnehmung sowie deren automatisierte Klassifikation und die Herleitung von Fahrsituationen bzw. Fahrszenarien für die virtuelle und reale Funktionserprobung im Simulator sowie im Fahrversuch.

In den Beiträgen zur künstlichen Intelligenz und deren Anwendung beim automatisierten Fahren werden die Kernaussagen des Übersichtsbeitrages weiter ausgeführt und erhärtet. Ausgehend von der Frage „Ist die künstliche Intelligenz die Lösung für alle Probleme?“ werden die Entwicklung der künstlichen Intelligenz und des maschinellen Lernens im Automotiveumfeld vorgestellt und anhand von Beispielen die Möglichkeiten sowie Herausforderungen beim Training und der Evaluation aufgezeigt.

Das Themengebiet der Nutzerakzeptanz wird über eine Einschätzung zum Vertrauen in das System beim automatisierten Fahren eröffnet und erstreckt sich im Weiteren über Fragen zur Überforderung sowie Unterforderung auch im Kontext der Fahreraufmerksamkeit und einer fahrfremden Tätigkeit. Geschehen die fahrfremden Tätigkeiten beispielsweise auf dem Bildschirm des Kombiinstruments, so ist die Aufmerksamkeit des Fahrers zur Übernahme der Fahraufgabe durch Um- bzw. Abschalten des Bildschirms einfacher zu erreichen. Weitere Aspekte im Kontext der Nutzerakzeptanz ergeben sich durch den Fahrer (Mensch) in der Interaktion mit dem Fahrzeug (Maschine) sowie mit den damit verbundenen Interdependenzen.

Das Prozesse im Umfeld des automatisierten Fahrens zu überdenken und gegebenenfalls neu zu gestalten sind, unterstreichen die Themengebiete Versicherungsvisionen und Haftungsfragen auf der Fachtagung.

Die ATZ-Fachtagung Automatisiertes Fahren 2019 dokumentiert den Stand der Technik im oben aufgezeigten Themenspektrum aus Sicht der Automobilhersteller, Zulieferer und Wissenschaft. Die diskutierten Beiträge führen zu Antworten und neuen Fragen, die einerseits grundsätzlich und andererseits vertiefend in der Sache sind, und damit die zukünftige Entwicklung mit beeinflussen.

Die Fachtagung Automatisiertes Fahren 2019 zeigt wichtige Ergebnisse auf dem Weg zum automatisierten Fahren, die in den einzelnen Vortragsmanuskripten oder Vortragsfolien in diesem Tagungsband weiter ausgeführt sind. Der vorliegende Band enthält die Beiträge, die zur Veröffentlichung freigegeben worden sind.

Univ.-Prof. Dr.-Ing. Prof. h.c. Dr. h.c. Torsten Bertram
Technische Universität Dortmund
Wissenschaftliche Leitung der Tagung

Inhaltsverzeichnis

Autorenverzeichnis

Rico Auerswald Fraunhofer-Institut für Verkehrs- und Infrastruktursysteme IVI, Dresden, Deutschland

Gerald-Alexander Beese KTI GmbH & Co.KG, Lohfelden, Deutschland

Univ.-Prof. Dr.-Ing. Prof. h.c. Dr. h.c Torsten Bertram Technische Universität Dortmund, Dortmund, Deutschland

Hans-Joachim Bieg Robert Bosch GmbH, Stuttgart, Deutschland

Marcel Borrack AZT Automotive GmbH – Allianz Zentrum für Technik, Ismaning, Deutschland

Dr. Matthias Butenuth IAV GmbH, Chemnitz, Deutschland

Stefan Dreiseitel Continental Teves AG & Co. oHG, Frankfurt/Main, Deutschland

Alfred Eckert Continental Teves AG & Co. oHG, Frankfurt/Main, Deutschland

Dr. Philipp Ehring GH-Legal, Saarbrücken, Deutschland

Matthias Sebastian Fischer Institut für Konstruktionstechnik und technisches Design (IKTD), Stuttgart, Deutschland

Dr. Robert Fuchs JTEKT Corporation, Nara, Japan

Dr. Martin Grießer Continental Teves AG & Co. oHG, Frankfurt/Main, Deutschland

Dr. Johann Gwehenberger AZT Automotive GmbH – Allianz Zentrum für Technik, Ismaning, Deutschland

Dr. Jürgen Häring ETAS GmbH, Stuttgart, Deutschland

Martin Herrmann IPG Automotive GmbH, Karlsruhe, Deutschland

Marcus Jenke Institut für Konstruktionstechnik und technisches Design (IKTD), Stuttgart, Deutschland

Dr. Alexander Jungmann IAV GmbH, Chemnitz, Deutschland

Dr. Roland Kallweit IAV GmbH, Chemnitz, Deutschland

Dr.-Ing. Martin Keller ZF Group, Düsseldorf, Deutschland

Maged Khalil Continental Teves AG & Co. oHG, Frankfurt/Main, Deutschland

Helge Kiebach KTI GmbH & Co. KG, Lohfelden, Deutschland

Melanie Kreutner AZT Automotive GmbH – Allianz Zentrum für Technik, Ismaning, Deutschland

Dr. Christian Lang IAV GmbH, Chemnitz, Deutschland

Paula Lassmann Institut für Konstruktionstechnik und technisches Design (IKTD), Stuttgart, Deutschland

Dr. Christoph Lauterwasser AZT Automotive GmbH – Allianz Zentrum für Technik, Ismaning, Deutschland

Christian Lienke Technische Universität Dortmund, Dortmund, Deutschland

Qi Liu BFFT Gesellschaft für Fahrzeugtechnik mbH, Gaimersheim, Deutschland

Thomas Maier Institut für Konstruktionstechnik und technisches Design (IKTD), Stuttgart, Deutschland

Andreas Mank Electrobit Automotive GmbH, Böblingen, Deutschland

Dr. Carlo Manna ZF Group, Düsseldorf, Deutschland

Dr. Stefan Milz Valeo Schalter und Sensoren GmbH, Kronach, Deutschland

Dr. Ralf Münzenberger INCHRON GmbH, Potsdam, Deutschland

Tomohiro Nakade JTEKT Corporation, Nara, Japan

Dr. rer. nat. Till Nattermann ZF Group, Düsseldorf, Deutschland

Thomas Otto Fraunhofer-Institut für Verkehrs- und Infrastruktursysteme IVI, Dresden, Deutschland

Jens Petersohn Elektrobit Automotive GmbH, Böblingen, Deutschland

Raphael Pfeffer IPG Automotive GmbH, Karlsruhe, Deutschland

Dr. Florian Pinsker IAV GmbH, Chemnitz, Deutschland

Fabian Pütz TH Köln, Köln, Deutschland

Dr. Michael Reichel Elektrobit Automotive GmbH, Böblingen, Deutschland

Florian Reichelt Robert Bosch GmbH, Stuttgart, Deutschland

Carsten Reinkemeyer AZT Automotive GmbH - Allianz Zentrum für Technik, Ismaning, Deutschland

Olaf Schmidt INCHRON GmbH, Potsdam, Deutschland

Manuel Schmidt Technische Universität Dortmund, Dortmund

Markus Schneider Continental Teves AG & Co. oHG, Frankfurt/Main, Deutschland

Thomas Schöpfner ETAS GmbH, Stuttgart, Deutschland

Jörg Schrepfer Valeo Schalter und Sensoren GmbH, Kronach, Deutschland

Naoki Shoji JTEKT Corporation, Nara, Japan

Mirko Taubenreuther IAV GmbH, Chemnitz, Deutschland

Simon Tiedemann Electrobit Automotive GmbH, Böblingen, Deutschland

Gregory-Jamie Tuezuen Institut für Konstruktionstechnik und technisches Design (IKTD), Stuttgart, Deutschland

Julius von Falkenhausen BFFT Gesellschaft für Fahrzeugtechnik mbH, Gaimersheim, Deutschland

Johannes Wagner ETAS GmbH, Stuttgart, Deutschland

Christian Wissing ZF Group, Düsseldorf, Deutschland

Mitsuko Yoshida JTEKT Corporation, Nara, Japan

Validation of level 4-5 functions with a cloud-based simulation

Jürgen Häring, Johannes Wagner

T. Bertram (Hrsg.), *Automatisiertes Fahren 2019*, Proceedings,
https://doi.org/10.1007/978-3-658-27990-5_1

Validation and verification of Level 4-5 Functions

The validation and verification of level 4-5 highly automated driving (HAD) functions requires a vast number of test kilometers. If real test drives are applied exclusively, it can be safely assumed that the required test scope cannot be covered within feasible limits of time and money [Winner]. This is not only due to the functional complexity of HAD applications, but also due to short development cycles and legal safety requirements.

In addition, the deployment and activation of new vehicle functions is desirable even after the regular "start of production" (SOP). In case of HAD, each of these new functions must fulfill the high demands on functionality and safety of automated driving functions. Validation and verification will thus not end with the regular SOP in the future - it has to be repeatable after SOP as well.

Caused by the high complexity of the automated driving task, HAD functionality is typically implemented by distributed systems consisting of several subsystems, such as ECUs, sensors, and buses. Consequently, validation and verification of the entire system must consider the interaction and communication of these subsystems.

All these factors indicate that intelligent validation and verification strategies are required for HAD testing [Lattemann] [Schöner] [Madrigal], which employ different methods and their combinations, including real driving tests, hardware-in-the-loop (HiL), and software-in-the-loop (SiL) testing [Wagner]. Real vehicle tests or hardware-in-the-loop tests will always stay indispensable as long as real components need to be considered in the test scope. Nonetheless, the validation and verification strategies will remain manageable only, if the majority of testing tasks is transferred from real vehicles on the road to virtual systems. These systems also help to ensure correct system functionality during dangerous driving maneuvers, for situations at the physical limits of driving dynamics, and during accidents – all of which would be impossible to investigate in real test drives.

The required amount of testing clearly suggests the usage of a technology that allows large-scale parallel operation of virtualized systems. It is common understanding that engineers cannot complete the required validation and verification tasks without such a technology. Consequently, cloud computing is a key to master the challenges of HAD validation and verification, because it allows transferring the testing tasks from the road to simulations on a large scale.

Most existing tools in the area of automotive system development and test are designed for the execution on single computers. In the case of HiL setups, they even run on dedicated real-time systems. This legacy inhibits the fluent transition of existing tools and applications into the cloud and prevents companies from taking benefit of cloud technologies. Hence, one key to cloud-based HAD validation and verification is making the existing, growing and cost efficient cloud resources accessible and available to automotive engineers.

Large Scale SiL Simulations for AD

The present paper focuses on the SiL methodology as a core element of HAD validation and verification strategies. The SiL methodology couples ECU target code, as it is implemented in the real vehicle, with virtual system components and simulation models of the vehicle and its environment. It allows investigating the overall system behavior in virtual test runs, providing full reproducibility, the ability to perform safety critical tests without risks, the test of the system reaction when system components fail (failure injections), system tests under heavy load on the automotive buses, validation of the system behavior in critical driving scenarios, and automated search for critical driving scenarios. These are only some examples for the numerous advantages of SiL based validation and verification.

SiL testing can be applied to different architectural levels of vehicle subsystems. It supports test and validation on the level of individual ECU functions (e.g. the HAD function planning layer), on the level of whole ECUs (e.g. an ECU implementing the driving functionality), or on the network level of multiple ECUs implementing the full autonomous driving function (e.g. integration of perception-, driving function system-, brake- and steering-ECU). While the focus of this paper is to discuss the test of ECU networks, it implicitly covers the other, less complex use-cases as well.

A SiL system for HAD validation and verification needs to reflect the specific requirements that driving automation implicate. E.g., the methodology has to be applied on a larger scale with respect to number of contributors to build a simulation, amount of scenarios or cases to be simulated, and the need to reuse the simulation environment for as many development steps as possible.

Modularity Based on Standards: Fostering Collaboration

The technical challenge of developing HAD vehicles can only be mastered by the cooperation of many parties. The overall task of designing and validating the autonomous vehicle thereby splits into smaller sub-tasks. Different parties take care of these sub-tasks, acting as component suppliers to the system integrator who builds the overall HAD vehicle. The system integrator also validates the integrated system. Similar to the integration of the real vehicle, the suppliers deliver a "digital twin" of the real components for the integration into the virtual vehicle, which is then used for the creation of a SiL validation and verification toolchain. As an example, virtual ECUs have to be provided by the ECU providers, sensor models are required from the sensor providers, and vehicle model (parameters) are needed from the vehicle chassis designers.

During the development cycle of an HAD vehicle a broad set of user groups is involved, having an even broader set of requirements to the SiL toolchain. Starting with the provision of the different individual system components over data driven open-loop testing to closed loop system integration and large scale parallel execution of driving manoeuvres according to a driving manoeuvre catalogue, the collaboration among different users, teams, and roles needs to be supported.

ETAS tools support these use-cases by consequently following a modular tool architecture based on standards. They can thereby be used as standalone installations, or they are integrated into customer specific toolchains using their powerful API interfaces.

Scalable SiL: From Desktop to Cloud Computing

Creating a SiL environment in cooperation with development partners is typically done incrementally, allowing to manage the efforts and the complexity that are associated with this task. It typically starts in a reduced setup, where a smaller user group integrates components in a SiL environment on a desktop computer. As soon as this set-up works successfully, it is enriched by more accurate parameterization and test-cases, then integrated in a larger virtual vehicle, and so on. When the number of required simulations or the complexity of the SiL simulation increases, e.g. by running virtual test drives over many millions of kilometers, the need for computational power is increased. To satisfy this need, scalable architectures are required, e.g. server clusters or a cloud infrastructure. They are ideally suited to support the demands for computation power of SiL based HAD validation and verification.

Implementation of a SiL Simulation Solution for HAD

As outlined before, the core idea of SiL validation and verification is to couple virtual ECUs *(vECU)* with *plant models*, i.e. a virtual car and its environment. The resulting system simulation is deployed to a *simulation target* to observe the functional behaviour of the objects under test in a virtual environment. Simulation and coupling of plant and vECU is done using an *integration and co-simulation tool*. Since ECUs are connected via automotive busses in the real car, the *simulation of automotive busses* is also required for the overall SiL solution. This “SiL simulation core” can be completed by additional tools, e.g. by test automation tooling to enable continuous testing use cases.

ETAS provides all the artefacts and supportive tools required to create such a “SiL simulation core”. In the following, we further focus on the integration and simulation platform as well as on tools to generate virtual ECUs. They all have in common that they support the previously described requirements on collaboration and scalability. At the same time, they can be flexibly deployed from local user PCs to servers and cloud infrastructures to support the various setups for HAD validation and verification as outlined above.

Integration and co-simulation platform

The ETAS COSYM integration and simulation platform allows users to integrate virtual ECUs, software and plant model artefacts to simulate the (virtual) HAD vehicle with its environment. It allows users to integrate and connect simulation models (e.g. as FMUs) and virtual ECUs built on different technologies. ETAS COSYM can deploy the simulation on various simulation targets. It supports real-time targets used for HiL testing, SiL simulation on a desktop computer, and the SiL simulation in the cloud. It can either be used as a full stand-alone installation on a desktop computer for interactive usage, or as a simulation backend service as part of a continuous integration toolchain.

Open and Modular Architecture

To support the need for cooperation as described above, ETAS COSYM follows a modular and open tool architecture with interfaces based on standards. This way, the tool can be tailored to the users' needs and be integrated into existing tool chains. As examples, COSYM allows the import of simulation models via the FMU standard, or the coupling to test automation tools using the ASAM standards.

Even more important, COSYM is fully controllable via an API and can thereby be integrated into continuous integration processes – including the execution of COSYM as a backend simulation service in the cloud.

From Desktop to Cloud to Real-time Computing

By its ability to deploy the simulation run into the cloud, ETAS COSYM also allows users to take advantage of the full, highly parallel computation power of the cloud. The simulation results exactly stay the same, independent if the cloud or a desktop computer is used. ETAS COSYM is designed to operate with all major commercial cloud infrastructure providers or, alternatively, on in-house cloud solutions. The user can take advantage from the vivid market of cloud providers, originating from IT dominated applications and offering computational resources at increasingly cheaper prices compared to the costs of operating own computation infrastructures.

Finally, ETAS COSYM also supports to deploy the simulation a Hardware-in-the-Loop (HiL) system operating in real-time. This allows a maximum re-use of artifacts created for the SiL simulation in the HiL based validation steps.

Simulation of Automotive Busses

Autonomous driving functionality is implemented distributed over a network of ECUs. The reasons for this are manifold, e.g. the reuse of legacy from existing EE architectures, modularity to cover system variants, the need for specialization of controllers on one task (e.g. brake controller with specialized circuits to control hydraulic systems), or safety considerations. In the vehicle, this network of ECUs consists of automotive communication buses. In a similar manner, simulated automotive buses are used to integrate virtual ECUs into a virtual vehicle in a SiL simulation. SiL environments therefore also need to provide the capability to simulate automotive bus functionality.

ETAS provides the simulation of automotive busses as part of the integration and simulation tool chain COSYM. This way, numerous system properties of the HAD vehicle can be investigated, e.g. the system behavior under the influence of communication latencies induced by heavy bus loads; the performance of diagnosis algorithms identifying failures in the bus communication; and the system stability if whole ECUs fail and backup strategies in the HAD vehicle are applied. In all these cases, the properties of the inter-ECU communication change dynamically. With COSYM, they can be investigated, since bus communication is an integral part of the overall SiL setup.

Virtual ECUs

To handle both safety requirements as well as functional requirements equally well, hybrid architectures are often used for HAD ECUs. They contain at least one microcontroller, which typically features a software architecture based on the AUTOSAR Classic Platform standard. In addition to that, one or multiple microprocessors take care of high performance calculation tasks. AUTOSAR proposes a software architecture for these microprocessors as well, which is described by the AUTOSAR Adaptive Platform standard [Bechter].

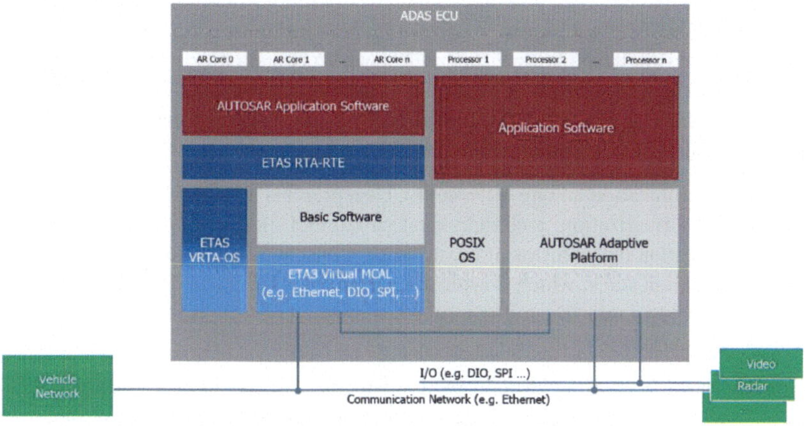

Figure: Possible Heterogeneous HAD ECU Architecture

Virtualizing an ECU, which is based on such a heterogeneous architecture, requires separate consideration of both, microcontrollers and microprocessors, which have fundamentally different properties. The AUTOSAR Adaptive Platform proposes a POSIX based architecture for the microprocessors. For their virtualization, off-the-shelf solutions exist. They are not discussed here.

The virtualization of the microcontroller imposes bigger challenges. Microcontroller architectures are substantially different from PC or server-based architectures. Yet ECU virtualization requires the execution of code, which is intended for a microcontroller-based target system, on a PC or a server. ETAS provides a solution, which allows bridging these differences while at the same time maintaining a maximum possible congruence between the behavior of the virtualized ECU and the real ECU, including aspects of scheduling, multi-core operation, and others: the ISOLAR-EVE product.

ISOLAR-EVE distinguishes carefully between hardware-independent and hardware specific code. All code that is independent from the target hardware platform, i.e. application software (ASW), the AUTOSAR runtime environment (RTE), and large parts of the AUTOSAR basic software (BSW), will be directly reused for the virtual ECU without changes. The remaining code, i.e. the code that is hardware specific, will be replaced with implementations for the PC that provide comparable functionality. In particular, there are specific implementations of the operating system (OS) and the microcontroller abstraction layer (MCAL), which substitute the corresponding microcontroller-specific implementations. While the OS is based on ETAS' production OS implementation that has been proven in hundreds of millions of production vehicles, the MCAL is specifically designed to use the PC platform in an optimal way. ISOLAR-EVE hence allows building virtual ECUs, which contain nearly the complete functionality of real production ECUs and thus reproduce their behavior as realistically as possible. This is because the virtual ECUs contain not only the ASW, but also nearly the complete BSW and large parts of the production OS.

Since the communication among different ECUs represents an essential aspect of the overall system virtualization, each individual virtual ECU needs to support the relevant communication mechanisms of the network. Consequently, including the respective communication stack, which is typically part of the AUTOSAR BSW, in the ECU virtualization, and an interface to the virtual bus implementation, is indispensable for a realistic SiL setup.

Depending on the goal of the virtualization, the required degree of detail may vary from ECU to ECU. In some cases, it may be sufficient to reproduce an ECU's functional aspects without reflecting all details of network communication. ISOLAR-EVE supports functional validation on application software level as well as the verification of individual software components, integration testing, or regression testing on the complete ECU software. For this purpose, the corresponding virtual ECUs provide interfaces on different levels above the BSW, e.g. the Functional Mockup Interface (FMI) for signal exchange on the level of physical signals, or open interfaces for parametrization of the virtual ECU and the manipulation of ECU internal data.

Through its openness and modularity, ISOLAR-EVE ideally supports the integration into continuous integration toolchains. It allows the creation of virtual ECUs for the local PC at the developer's desk over server clusters and cloud architectures to real-time setups. It goes without saying that the integration of virtual ECUs with the ETAS COSYM co-simulation platform and with virtual bus networks is supported as well.

Obviously, a standardized SiL architecture, as mentioned above, will also play an important role for the interfaces of virtual ECUs in the future.

Application of Technology

ETAS participates in a project, in which a large scale, cloud based SiL simulation environment is created supporting the validation of an ECU and its parameterization in virtual endurance runs. The project implements a graphical user interface (GUI) that allows automotive engineers to configure virtual test drives for the validation of virtual

ECUs. The GUI is fully customized to support the workflow and needs of the engineer using the tool. By the click of a button, the engineer can start the virtual test drive. In the background, the tool deploys the SiL simulation to a cloud infrastructure, where, e.g., 2000 parallel simulation jobs are started, executed, and the results are provided to the GUI front-end. The massive parallelization leads to the situation that the simulation time is not limited by the duration of the virtual test drive, but by the general time to start, deploy and finish the simulation – summing up to only a few seconds. The massive amount of simulation data remains in the cloud to perform the data analysis step there.

The modular structure of ETAS COSYM is used for a strict separation of simulation backend service and user interface. The cloud and the IT technology used is hidden to the user who can focus on the analysis of the simulation results. He does not have to care about the technical hurdles of operating a cloud simulation. In later stages, the fully scriptable simulation service will be integrated in continuous integration & test toolchains.

The project also takes advantage from the consequent design of ETAS COSYM for cloud operation. The tools generates code consuming a minimum amount of computational resources only. This is important since cloud payment models are based on resource usage. ETAS COSYM is designed to minimize the costs of operation in the cloud.

Summary

The development of HAD vehicles demands for cloud based SiL validation and verification. By this, the need for cooperation between many parties and the demand for high computational power can be satisfied. To address this, the simulation and integration platform ETAS COSYM and ISOLAR-EVE are designed consequently to operate in the cloud. The use of micro services enables continuous integration and testing approaches, cooperation among multiple engineers working at different locations, and flexible usage of the co-simulation platform. Since COSYM supports the most important cloud infrastructure providers on the market, companies are free to choose the platform of their choice and take advantage of the dropping prices in the competitive market of cloud infrastructure providers.

From the viewpoint of the HAD vehicle development engineer, ETAS tools pave the way to take benefit of the rapidly developing cloud technology. The developers can focus on their core tasks, e.g. system design and function development, rather than handling technical hurdles of cloud operation.

References

[Bechter] Dr.-Ing. Markus Bechter, Dr. Marcel Wille, „AUTOSAR for next generation cars - Smooth interaction of different software platforms", https://www.vdi-wissensforum.de/news/autosar-for-next-generation-cars/

[Lattemann] F. Lattemann, "Approaches to Test L4+ Urban Automated Driving Systems", ETAS Connectinos Conference, Stuttgart, 2018

[Madrigal] Alexis C. Madrigal, „Inside Waymo's Secret World for Training Self-Driving Cars", https://www.theatlantic.com/.../inside-waymos-secret.../537648/, 2017

[Schöner] H.-P. Schöner, "The Role of Simulation in Development and Testing of Autonomous Vehicles", Driving Simulation Conference, Stuttgart, 2017

[Wagner] Johannes Wagner, Joachim Löchner, Dr. Oliver Kust, "Virtualization for Verifying Functional Safety of Highly Automated Driving Using the Example of a Real ECU Project", 4th International ATZ Conference on Automated Driving, Wiesbaden, 2018

[Winner] Hermann Winner, "Quo vadis, FAS?", ATZ/MTZ-Fachbuch, Handbuch Fahrerassistenzsysteme, Wiesbaden: Springer Vieweg, 2015

Handling complex systems: systems design for AD vehicles

Martin Grießer, Maged Khalil and Stefan Dreiseitel

Systems & Technology, Continental Chassis & Safety

T. Bertram (Hrsg.), *Automatisiertes Fahren 2019*, Proceedings,
https://doi.org/10.1007/978-3-658-27990-5_2

Content

Continental

Digitalization requires new Vehicle Capabilities

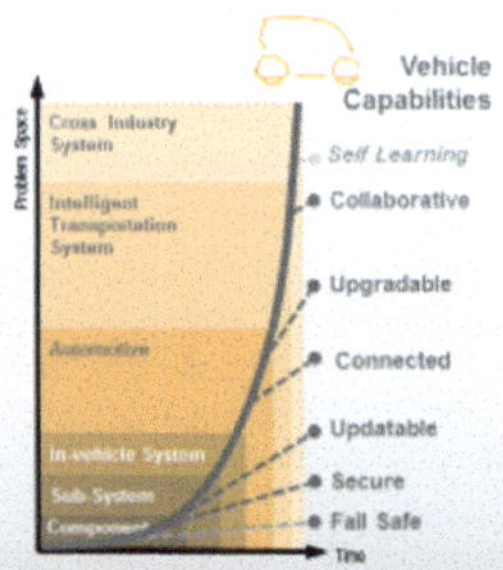

Digitalization challenge

- The vehicles of today learn ONCE
- The vehicles of tomorrow will learn over their life time

Key driver

- automated driving
- electrification
- integration of the environment
- time to market of new features

Continental

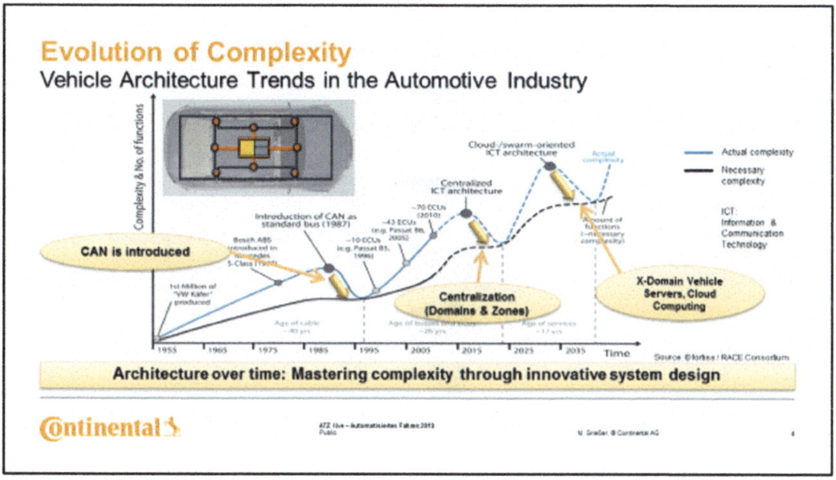
Evolution of Complexity
Vehicle Architecture Trends in the Automotive Industry
Complexity & No. of functions
Cloud-/swarm-oriented ICT architecture
Actual complexity
Necessary complexity
ICT: Information & Communication Technology
Centralized ICT architecture
Introduction of CAN as standard bus (1987)
CAN is introduced
Centralization (Domains & Zones)
X-Domain Vehicle Servers, Cloud Computing
1955
1965
1975
1985
1995
2005
2015
2025
2035
Time
Architecture over time: Mastering complexity through innovative system design
Continental

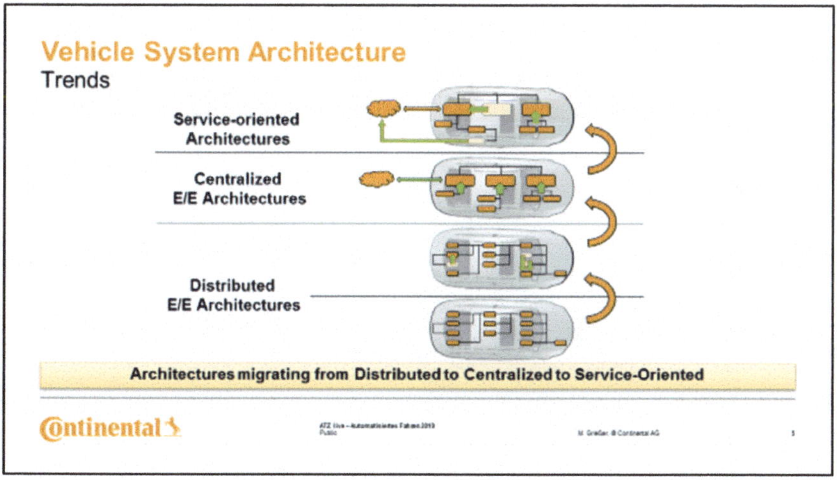
Vehicle System Architecture
Trends
Service-oriented Architectures
Centralized E/E Architectures
Distributed E/E Architectures
Architectures migrating from Distributed to Centralized to Service-Oriented
Continental

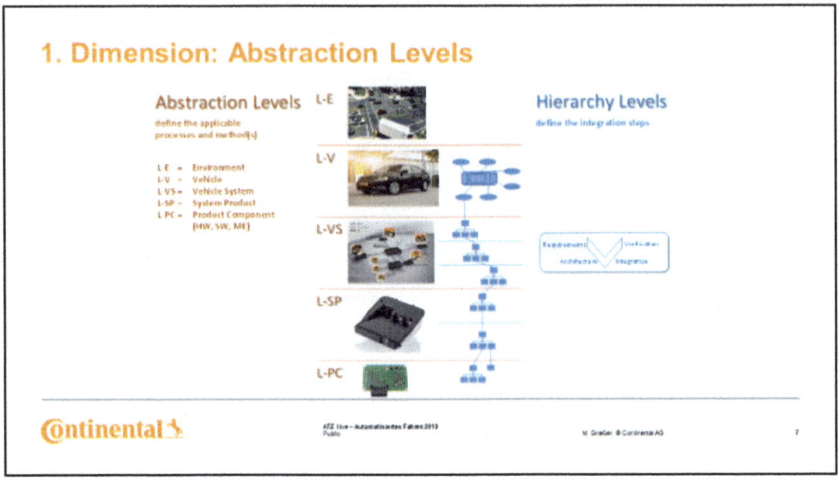
1. Dimension: Abstraction Levels
Abstraction Levels
define the applicable processes and method(s)
L-E = Environment
L-V = Vehicle
L-VS = Vehicle System
L-SP = System Product
L-PC = Product Component (HW, SW, ME)
L-E
L-V
L-VS
L-SP
L-PC
Hierarchy Levels
define the integration steps

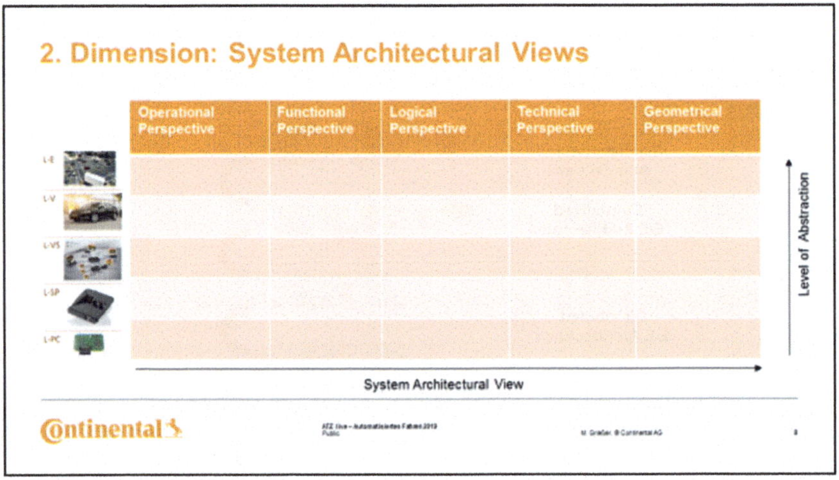
2. Dimension: System Architectural Views
Operational Perspective
Functional Perspective
Logical Perspective
Technical Perspective
Geometrical Perspective
L-E
L-V
L-VS
L-SP
L-PC
Level of Abstraction
System Architectural View

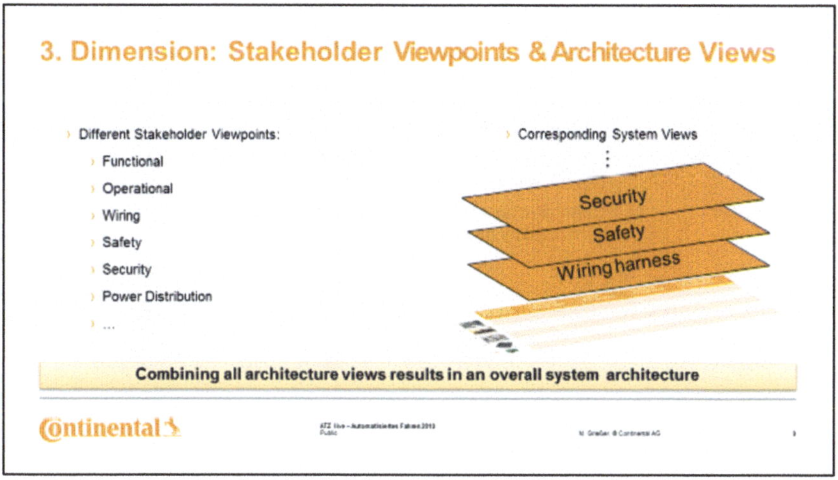
3. Dimension: Stakeholder Viewpoints & Architecture Views
Different Stakeholder Viewpoints:
Functional
Operational
Wiring
Safety
Security
Power Distribution
...
Corresponding System Views
Security
Safety
Wiring harness
Combining all architecture views results in an overall system architecture
Continental

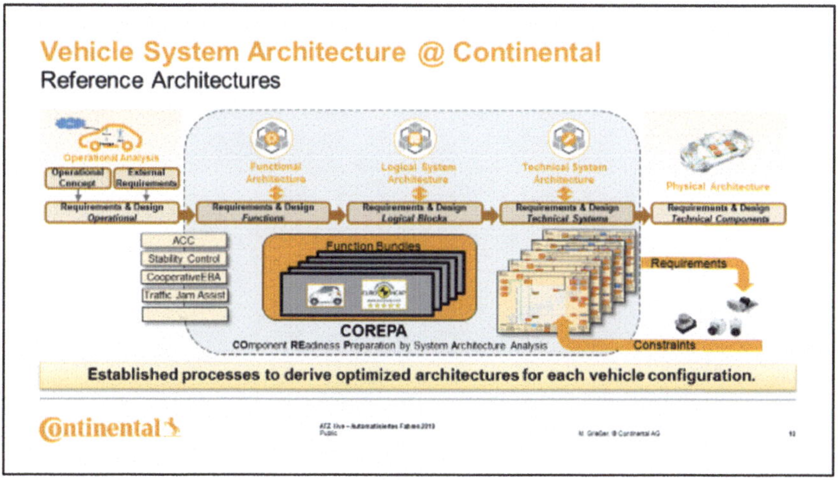
Vehicle System Architecture @ Continental
Reference Architectures
Operational Analysis
Operational Concept
External Requirements
Requirements & Design Operational
Functional Architecture
Requirements & Design Functions
Logical System Architecture
Requirements & Design Logical Blocks
Technical System Architecture
Requirements & Design Technical Systems
Physical Architecture
Requirements & Design Technical Components
ACC
Stability Control
CooperativeEBA
Traffic Jam Assist
Function Bundles
COREPA
COmponent REadiness Preparation by System Architecture Analysis
Requirements
Constraints
Established processes to derive optimized architectures for each vehicle configuration.
Continental

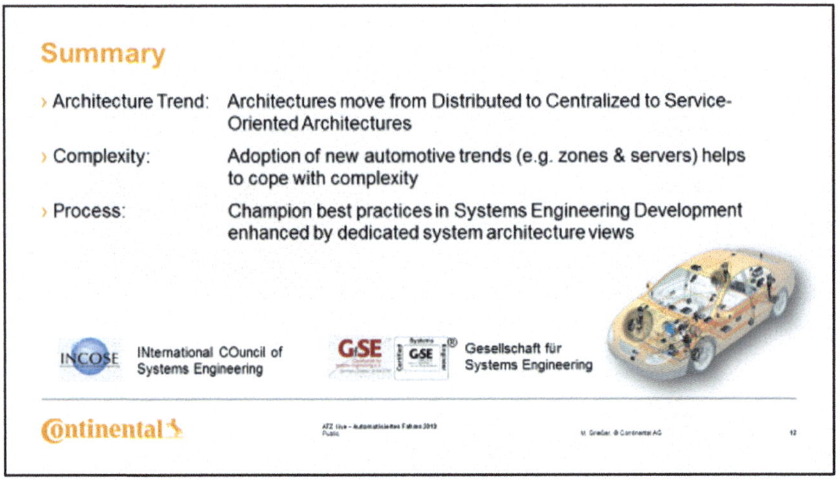
Summary
› Architecture Trend: Architectures move from Distributed to Centralized to Service-Oriented Architectures
› Complexity: Adoption of new automotive trends (e.g. zones & servers) helps to cope with complexity
› Process: Champion best practices in Systems Engineering Development enhanced by dedicated system architecture views
INCOSE
INternational COuncil of Systems Engineering
GfSE
Gesellschaft für Systems Engineering
Continental

HAD development game changers and changing SW creation

Dr. Michael Reichel

Jens Petersohn

T. Bertram (Hrsg.), *Automatisiertes Fahren 2019*, Proceedings,
https://doi.org/10.1007/978-3-658-27990-5_3

Introduction

Bringing automated driving functions and systems on the road is a highly complex task. We expect SW with >20 million lines of code per high performance computer and a significant increase in complexity especially for HAD applications. From the point of view of a software Tier 1 company, there are major game changers when it comes to developing such functions. These game changers will exert influence on the industry, including the actual way of working for everyone involved.

The ideas shared within this paper are likely to encourage discussions on management level on how automated driving and future mobility can be realized, both cost-efficiently and timely.

At a glance, we see six major game changers for Automated driving SW creation:

- Market game changer: SAE level increase
- Technology game changer: Artificial intelligence
- Technology game changer: Server based E/E and service oriented base SW
- Technology game changer: Sensor technology
- Process game changer: Feature increase beyond SOP
- Business game changer: Subscription models for driving features

This article will describe these game changers in more detail and reason their selection. The changes are reflected in these three major changes in SW creation which will be discussed:

- Invest post SOP and scale by usage
- Decoupled lifecycle of software and vehicle
- Data management as a key enabler for data reuse

Game changers

Market game changer: SAE level increase

One of the driving forces for change in the automotive industry is the aim to build and deploy (fully) automated driving vehicles. Premium OEMs such as AUDI, BMW or Daimler strive to offer their customers the best automation experience in the market and to be the first to allow conditional driving (SAE level 3) without the necessity to supervise the system[1]. These market participants focus on the market of mostly privately-owned

1 Sources: https://www.golem.de/news/level-5-bmw-will-fahrer-ab-2021-ueberfluessig-machen-wenn-er-will-1703-126780.html, https://www.audi-mediacenter.com/en/automated-driving-3651, https://techcrunch.com/2017/04/04/daimler-and-bosch-fully-autonomous-cars-within-5-years/

cars for personal mobility. On the other side of the spectrum, tech backed companies such as Waymo target for level 5 automation vehicles to allow new forms of mobility.

Figure 1 show one of the latest forecasts for market introductions and distributions of automated driving applications. Although these numbers are much lower than a year ago and reflect the stage change in the hype cycle (moving from the peak of inflated expectations towards the trough of disillusionment), the numbers indicate a continuous development towards higher level of automation.

The challenge has begun and has raised the interest of end customers. All of the following game changers stated in this paper are subsequent changes in order to fulfill the expectations.

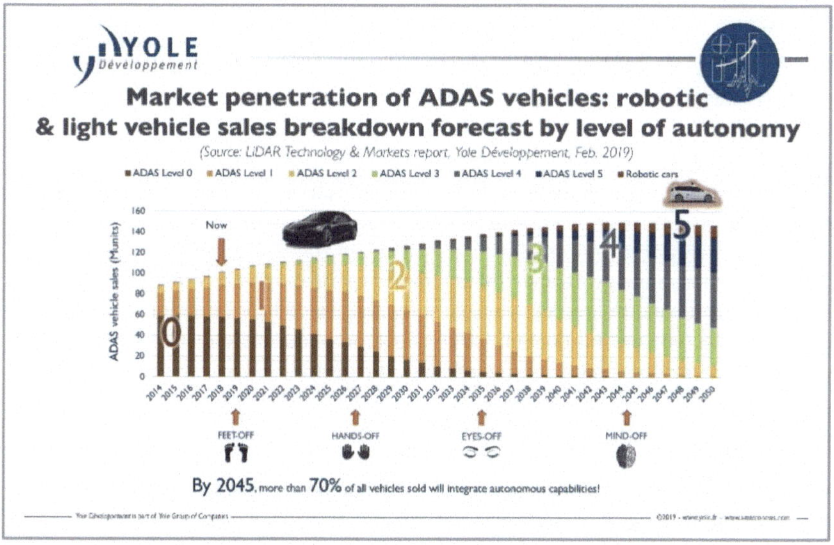

Figure 1: Expected increase of automated driving features, source: "ADAS Sensor Market & Technology Trends", Cédric Malaquin, International VDI Conference – Automotive Sensor Systems, February 2019

Technology game changer: Artificial intelligence

The usage of deep neural networks is seen as one – if not the only – technical solution to cover a wide scope of real-life scenarios for automated driving at SAE level 3 and higher. Although artificial intelligence ("AI") itself is not limited to only DNNs, the automotive industry refers to that term for data driven learning technologies such as DNNs.[2]

The major change in terms of SW development is the need for enough training data beforehand. While "old school" SW algorithm development could start using a very limited amount of recorded data and later utilize a significant amount of data to enable testing and validation – the application of AI methods changes the workflow significantly. This means

1. that whole sensor setup needs to be ready
2. a fleet of cars need to be equipped for data recording
3. data must be uploaded
4. meta data needs to be added to enable indexing, searching
5. reference = training target data needs to be added = labeling

Not only the amount but also the quality and the coverage of the data is of importance, analysis tools for data handling will become one of the most critical tools for software development. [3]

Technology game changer: E/E architecture and service oriented base SW

One major change in development of any in-car application will come with changes of E/E architecture and the respective base software. As depicted in figure 2, current architectures include central domain ECUs. Such architectures already bundle applications of the same domain on non-interchangeable central ECUs.

Looking into the future of E/E architectures, there are reasons for a further centralization and abstraction:

- easier integration of new features from continuous development
- safety applications need fallbacks starting at E/E systems level
- a reduction of ECUs results in BOM cost reductions and eases packaging
- using standardized and easy scalable central ECUs is the enabler for complex functions such as automated driving
- Hardware integration and E/E design become manageable for much more complex systems

---

2 S.M. Grigorescu, M. Glaab and J. Schlosser, "KI für Selbstfahrende Autos", EE Faszination Elektronic, http://www.industr.com/2303747/121541, 2017

3 Please see also Elektrobit's second presentation at this conference: "Solving the validation challenge of automated driving with a holistic test center."

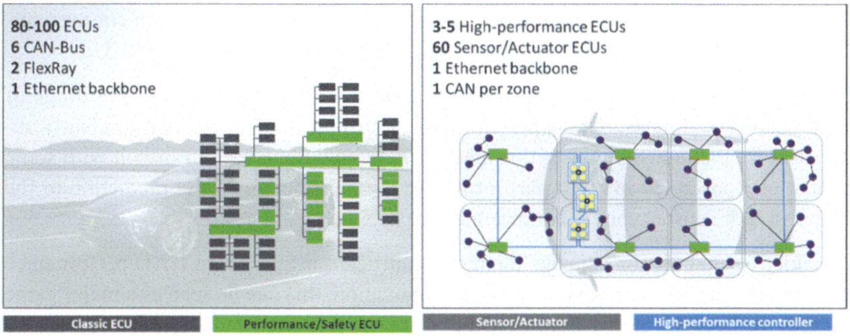

Figure 2: Depiction of E/E architectures, on the left evolving domain architecture, on the right future zonal architecture

There are also drawbacks and technical challenges to bring these advantages onto the road. Software integration will become much more complex, variant handling will shift from HW to SW variant management and testing of zonal based architectures will require new tooling. Nevertheless, we expect such architectures to be introduced in the future. These will come with service oriented base SW for further abstraction of the underlying hardware. SW architectures can rely on publish/subscribe technology rather than fixed connections. The base SW will ensure proper timings for critical paths of computation, will bring security and safety mechanisms and allow for easy transfer of applications in between the central backbone.[4]

Technology game changer: Sensor technology

Information about the vehicle´s surroundings, its dynamic motion state and global position as well as information about the passengers is the essential base for automated driving applications. Therefore, the questions of which sensors are used, where are they placed around and in the vehicle and what data they need to provide are core questions during the conception phase of a HAD application.

Looking at the current innovation density in the field of sensor technology one needs to state that the answer on which sensor will be used is quite a moving target. Next to the evolution of existing sensors and the innovation efforts of established T1s, there is a number of startups working on automotive grade sensor technology. Quanergy,

4 Please refer also to the presentation of Mr. Martin Schleicher: "Scalable and flexible software-platform for high-performance ECUs", Elektronik im Kaftfahrzeug, June 19th 2018

Luminar and AEye for example are working on new laser and/or camera systems. Nvidia is promoting camera-based AI algorithms processed on a central ECU.

Furthermore, the answer on what the sensor set will look like is open. The usage of different physical measurement principles to comply with safety concepts is clearly understood by the industry. But if this means a camera dominated sensor belt with some radars, or if laser sensors will play a more important role; any valid prediction of the proportion of sensor types is not feasible today.

Considering that sensor data has the most significant influence on HAD performance plus sensor data will change due to sensor innovation, this can be seen as a major game changer without the ability to foresee the winner of this challenge.

Process game changer: HAD feature increase beyond SOP

Some of the current software-based applications and features for end users are already improving over the vehicle lifecycle and are being updated on a regular basis. The term vehicle lifecycle is depicted in figure 2 and from an end-customer point of view it reaches from SOP with the production of the first mass produced car, passes the point of EOP with the production of the last vehicle, the time where the OEM supports with spare parts and maintenance until the "after service life" of undefined length.

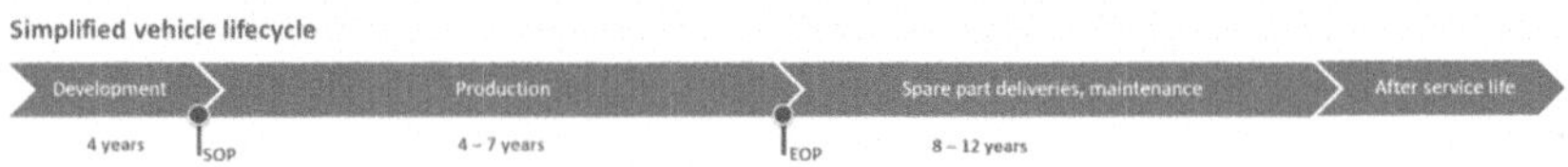

Figure 3: Simplified vehicle lifecycle

There are several examples of continuous feature increase along the lifecycle: Regular map updates which are being delivered more and more over the air[5] or software updates over-the-air for the automated driving features as Tesla is performing. Furthermore, Tesla is also aiming for over-the-air updates for ECU core SW.[6]

With these first examples of over-the-air updates one might ask: Why should an OEM decouple the lifecycle of the software and the lifecycle of the vehicle for HAD?

---

5 Source: https://news.ihsmarkit.com/press-release/automotive/over-air-software-updates-create-boon-automotive-market-ihs-says

6 Source: https://ihsmarkit.com/research-analysis/remote-software-update-future-growth-business.html

1. The development of automated driving as one of the most complex software packages might not be ready and would postpone a SOP.[7]

Learning technologies such as DNNs, also considered a game changer, and the concept of feature increase over the vehicle lifetime mutually reinforce each other: Learning technology require a lot of training data and it is simply impossible to collect all possible scenarios beforehand. The approach of selling vehicles with sensors to obtain the valuable data from customers and later introduce new features based on the received data appears to be a technical solution for this dilemma.

2. An OEM needs to keep his vehicle product attractive.

The attractiveness can easily be retained by enabling more and more roads to be driven automatically. Looking at other industries: Some "killer" features of mobile phones were not invented with the hardware, they were introduced in software independent of the hardware lifecycle.

3. An OEM can earn additional money with vehicles in the field.

Especially bigger OEMs have a significant number of vehicles in the field. Almost 10 Mio Volkswagen vehicles are on the road in Germany alone.[8] That opens a huge market for a software and service business.

Considering these possibilities and developments and extrapolating from the current rate of increasing features over lifetime, one can assume that the whole story of automated driving will be developed step by step and will hardly be finished by any specific SOP date.

Business game changer: Subscription models for driving features

A subscription business model is based on payments per use or per time period. Some features in the automotive industry are already commonly using a subscription model. BMW, AUDI, Daimler and VW offer connected features for a 1-3 year subscription.[9] There is no proof that this model will be successful, but if we assume that the other proposed game changers are valid, the industry will need continuous income streams to cover the revolving costs of continuous feature development, map and security updates. A subscription model could be a valid way to solve this issue. Furthermore, this model can bring automated driving functionality to a group of users that is not willing to pay

7 Example AUDI as source: https://www.golem.de/news/neuer-a8-vorgestellt-audis-staupilot-steckt-noch-im-zulassungsstau-1707-128881.html

8 Source: https://www.kba.de/DE/Statistik/Fahrzeuge/Bestand/b_jahresbilanz.html

9 Source: https://www.bmw-connecteddrive.de/app/index.html#/portal/store

a large premium upfront but can be convinced to subscribe for a monthly fee. Monthly payments for a full set of highly automated driving features can become as mainstream as paying to get access to millions of music records.

Derived changes in SW development

This paper focuses on three aspects of SW development that are about to change due to the game changers in technology, market and processes mentioned above:

- Invest post SOP and scale by usage
- Decoupled lifecycle of software and vehicle
- Data management as a key enabler for data reuse

Invest post SOP and scale by usage

Current HAD project business is usually calculated in several steps along the vehicle development process. Software is meant to be feature complete a year before SOP, and until the SW freeze shortly before SOP the bug fixing process is in focus. After sign-off only mandatory bug fixes, security or safety issues, are reasons to change the software again. With a fixed pricing on HAD features for the end customer at the time of purchasing, there is no business model that would enable continuous feature increase beyond SOP.[10]

With a decoupling of the vehicle and the software lifecycle and the introduction of subscription models, there is the possibility to build a continuous income stream from HAD feature usage. If the OEM can earn per use or per time, a whole industry will be able to invest after the SOP as well.

For software T1s that means a possible change from production piece-based or project earnings towards software release / subscription success-based earnings and hence investments.

The idea of such a continuous investment in feature increases can scale by all users and targets the market of all vehicles on the road using the platform software – independent of their model or brand.

10 See also: https://ihsmarkit.com/research-analysis/remote-software-update-future-growth-business.html

Decoupled lifecycle of software and vehicle

First estimations of how much software needs to be developed for stacks that are running on high performance computers indicate about 20 million lines of code for such systems. Compared to a high-end infotainment system (10 million), a high-end navigation system (4 million) or a typical body controller (up to 2 million) a SAE level 3 automated driving system will need significant resources for developing, integration and test. These numbers are one driving force for a decoupling of the software and vehicle lifecycle. Taking into account the game changers

- Technology game changer: Artificial intelligence

  The necessity of collecting data in mass production cars first and using the data to provide better functionality.

- Technology game changer: Server based E/E and service oriented base SW

  The ability to incorporate new functions into an existing architecture easily. The provision of enough abstraction to hardware and other software elements to be able to test incrementally.

- Process game changer: Feature increase over vehicle lifetime

  Customers are used to feature increases via software updates.

- Business game changer: Subscription models for driving features

  The OEM is enabled to earn additional money by providing new features to an existing fleet.

a decoupling of the lifecycles for software and the vehicle appears even more probable.

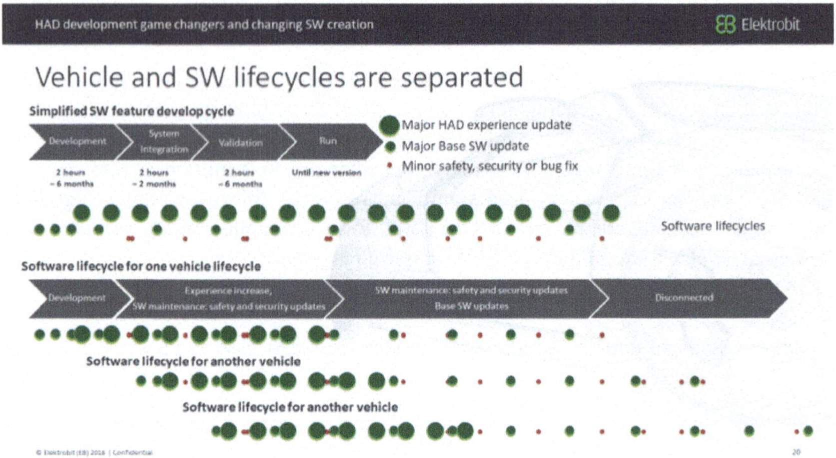

Figure 4: vehicle and software lifecycles

As soon as the decision to decouple the lifecycles has been taken, the setup to have basically one HAD lead development project independent of all vehicle segments comes into reach. OEMs introduced vehicle platforms to save money and increase re-use. Currently they use such platforms cross vehicle models and cross company brands. In respect to HAD Software a similar approach is a vehicle and brand independent software base development with continuous releases and updates.

A prerequisite is a fixed sensor setup or a few standard sensor setups across all brands and segments. By introducing such fixed setups, a continuous HAD development with releases in different brands and segments is possible.

Figure 4 depicts a simplified lifecycle model of in-vehicle software. It distinguishes base software (small green dots) and HAD feature software (big green dots). Base software is using open source (e.g. Linux) and hence already relying on a on a vehicle independent lifecycle. Such base SW is being updated on regular base either as minor update (red dots) or as new major version. This is a forward update strategy for all vehicles with the target to support and maintain only the latest base SW. With vehicle SW controlled by the OEM this is possible and necessary: Maintaining several branches for different cars (backward update strategy) would increase SW maintenance effort dramatically.

The HAD software is seen as application and has a different update cycle. It usually does not (yet) rely on open source parts and the lifecycle is defined by the OEM.

Data management as a key enabler for data reuse

Next to the development of software and hardware, there is another upcoming cost driver: The costs for data storage and management.

There are several use cases for recording and storage of large sensor data:

- Specific sensor data recording fleet with the latest sensors and reference systems
- Mass production vehicles recording with mass production sensor set
- Designated sign-off fleets to validate driving functions
- Simulations of test drives for testing or development that the user does not want to re-simulate on the fly

Only for the first use case one can expect 11 Mio USD per year, increasing over time for cloud data management costs. For this figure we considered that newly designed vehicle fleets bring roughly 50 cars on the road and each records up to 9 GByte/s of data (Camera belt, laser sensors, radar sensors, inertial sensor set, vehicle bus data and a redundant set of reference sensors). They drive 4 hours a day, 20 days per months. 5 % of the data is seen as hot data, rest is cold data. Costs cover the costs for uploading using datacenter access, meta data generation and cloud storage.

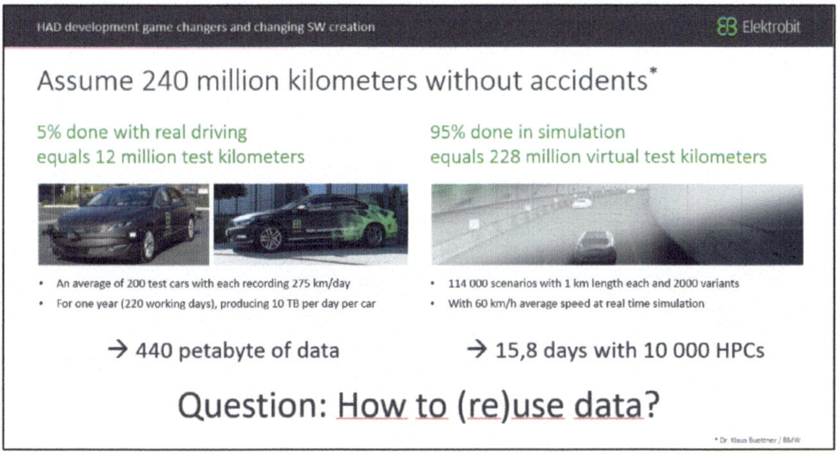

Figure 5: Data reuse as change result in HAD development

With the game changer mention in the first sections of this article, the amount of data is mandatory. Using and reusing the data becomes therefore very important.

Reuse is enabled by a fixed sensor sets of a longer time, the above-mentioned platform strategy for SW development and intelligent tooling to access and maintain data. Especially for this topic there is a separate talk of Simon Tiedemann of Elektrobit: "Solving the validation challenge of automated driving with a holistic test center".

Conclusion

This paper introduced six game changers for software development with the focus of automated driving software. Additionally, this article postulates three changes for software development.

One can argue that there are more or other game changers with a larger impact or that some of the conclusions will not come true in the future. If this happens, the goal of this paper has been reached: To encourage the industry to talk about these changes and possible futures needs. We will be more successful in the future with a common understanding of the future we want to create.

Start the flow! Why timing of autonomous driving functions needs to be taken into account at an early stage

Olaf Schmidt
Dr. Ralf Münzenberger

INCHRON GmbH

T. Bertram (Hrsg.), *Automatisiertes Fahren 2019*, Proceedings,
https://doi.org/10.1007/978-3-658-27990-5_4

Abstract

A large number of sensors and actuators, combination of high-performance computer and safety critical applications, complex data flows with a huge amount of data and many Tier-2 software suppliers are fundamental challenges for the development of autonomous driving systems. To handle the technical and organizational complexity the most proven development methods and tools are just good enough. Many customer projects have shown focussing early on timing is a key for success. Using state-of-the-art simulation and worst-case analysis tools for evaluation of real-time requirements and design optimization does massively reduce project costs and risks.

1 Timing in autonomous driving

In autonomous driving systems many components like sensors, algorithms and actuators have to work together seamlessly. It is our experience from working in ADAS for more than 10 years and autonomous driving for several years, that timing is crucial in almost every project due to the high complexity of the applications. Although there are established methods and tools in place timing is still not considered enough during the requirement, design, implementation and test phase. In this article we give an overview of a methodology developed and proven in many development projects that focusses on timing during the architectural design phase.

The key area of interest is the data flow between sensors and actuators. How and when is data created, communicated, processed and results are provided? If the data flow doesn't work correctly in some aspects (e.g. safety, reliability) the autonomous driving system will never get the official approval and homologation.

Because data flows are so important, we use the concept of event chains for the specification of data flows and their requirements successfully since many years. Event chains can be thought of as the path from an input, such as a sensor, through software elements, until the desired output, such as an actuator movement, occurs.

A big strength of using event chains is to create and to specify real-time requirements of data flows explicitly and precisely. A good example is an emergency breaking assist. The maximum end-to-end latency must be less than 300 ms from the point in time when an obstacle pops up, until the car starts breaking. The end-to-end event chain goes through 3-5 embedded control units (ECUs) and 4-6 buses.

In many cases implicit requirements lead to software that operates correctly in an isolated development environment, but not in a complex integrated system. This is a common source of error for rare violations.

In addition, event chains can be used as a communication tool between globally distributed teams in a company as well as Tier-2 suppliers and their integrators. A robust and scalable architecture, paired with early proactive anticipation of real-time issues, almost always delivers far superior results than just trying to fix real-time issues discovered by chance in later phases of the development lifecycle.

For us, excellence in real-time means to identify sporadic runtime issues systematically and right from the beginning, thus making the development of complex real-time systems predictable in terms of product quality, development schedule and costs.

2 Event chains

Looking from a 10,000 foot level data is generated by multiple sensors, processed by the ECU and the actuators must be controlled at the right point in time. The processing may include various steps of pre-processing, data fusion and post-processing using multiple parallel data flows on a modern multi-core SoC.

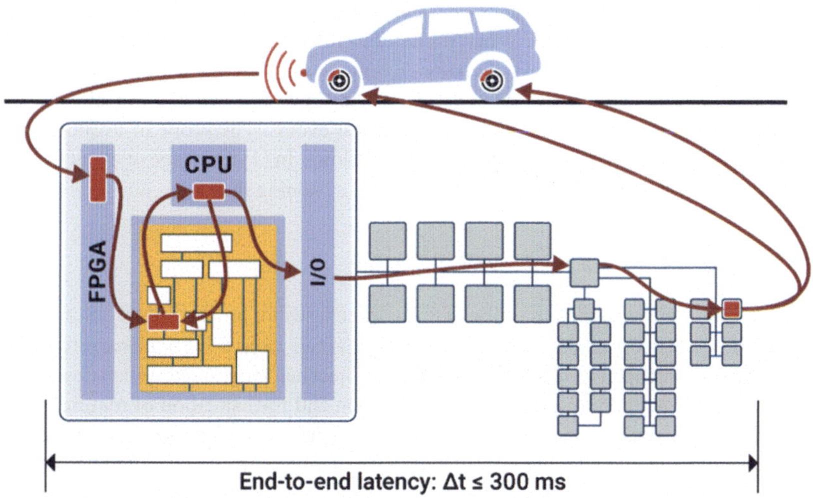

Figure 1: Event chain of an emergency brake assist

From the perspective of the customer function in the example above the data has to be processed within a certain timeframe from the radar to the brake. The real-time requirements can often be derived from physical properties (e.g. at a certain speed the vehicle has a certain amount of milliseconds to stop in front of an obstacle).

The processing of the data and the data transfer causes communication latencies and execution times of the software. Both depends on the amount of data, the bandwidth of the connectors, the complexity of the algorithms as well as bus arbitration and scheduling effects of the software.

In general, event chains are a temporal and causal order of events. That means an event (chain step) is causally dependent on its predecessors. A event chain step can be the point in time an obstacle occurs, sending a message on a bus, or calling the entry function of a software application. The concept of event chains is relatively easy to understand and very powerful, even if mastering the challenges is far from being simple.

In Figure 1 the data flow of an emergency breaking assist from the radar via several ECUs and busses to the brake is shown as event chain (red arrows). The end-to-end latency requirements of this even chain is 300 ms. So it is precisely specified which hardware, software and busses the requirements relates to. It is very important to have a common understanding between architects, software developers, integrators and the testers especially in collaboration projects with several companies.

For today's ADAS, about 50 time-critical event chains have to be jointly optimized to meet all real-time requirements. When it comes to fully autonomous driving, it will be well above 1000 time-critical event chains. Event chains are a good way to represent, design and test the data flow. The end-to-end real-time requirements can directly be associated with an event chain and its first and last events. The scope of event chains should be real-time critical and safety critical data flows. In [1] the usage of event chains on different architecture levels for a safety critical system is shown.

3 Requirements of event chains

Poor requirements management is one of the major causes for project failure. (Software) Engineering practice has also shown, that design issues caused by poor requirements are increasingly difficult to resolve the further a project has progressed. When designing new features for autonomous driving, the definition and formalization of requirements concerning the dynamics of event chains should be performed from the very beginning of the development process. The compliance with these requirements should be continuously verified by means of model-based simulation, formal analysis, and by tests performed on the real target platform [2].

For the evaluation of event chains, the following metrics have proven to be useful:

- *End-to-end latency*: in the context of embedded control systems, the end-to-end latency is defined as the time it takes for a signal detected by a sensor (source) to pass through the system and cause a response at the attached actuator target. In addition from the perspective of an ECU architect or integrator the scope of the end-to-end latency could be receiving input data from the bus and sending the output data on the bus. This means requirements may only apply for a part of the event chain.

- *Data consistency*: In general, the functional correctness and quality of automated driving applications depends on the precisely executed fusion (*synchronicity requirement*) of data from multiple sensor sources (SDF). Loss of data – causing an event chain breakup (*break-off requirement*) – or processing of inconsistent (outdated) data (*multiple processing requirement*) can lead to malfunctions. Inconsistencies are typically caused by misconfiguration or misbehavior of activation periods of processes involved in the sampling and (pre-) processing of sensor data. Corresponding real-time requirements, like start-to-start jitter or data age, are derived from the physical domain by application of sampling theory (e.g. Nyquist-Shannon sampling theorem) during the design of SDF algorithms.
- *Execution order:* The execution order of functions usually follows the data flow. However, the decomposition and allocation of functions to processes is also guided by safety and security constraints, which may require a process architecture design that conflicts with real-time requirements imposed by the functional domain.

But where do requirements come from? Naturally an OEM has to provide precise and comprehensive requirements for the overall systems and break them down to the subsystems. It is the responsibility of the Tier-1 to demand sufficient requirements from the OEM that are relevant for his subsystem. The Tier-1 has to break these requirements further down to the relevant parts of the subsystem like SoCs, cores, tasks and runnables/functions. Some requirements may also become an input to Tier-2 suppliers (e.g. software components).

Sometimes people confuse requirements with implementation details. For example, a specification like "my application has to be executed every third time after a certain task, but always before another task" is a possible implementation be not a requirement. Here it is the task of the requirements engineer to clearly mark the difference.

We have seen in many projects that typical real-time requirements like bus or CPU load, response-times of tasks or ISRs (interrupt service routines) are fulfilled, but the system is erroneous. If this happens, we'll have a look the event chains.

4 Model-based design, analysis and optimization

So, how can real-time requirements be verified early in the development process? How can I make sure, that the architecture fulfills the requirements, before I start the implementation?

As no real system is yet available and to handle complexity a model-based design approach is used. It is state of the art to use models for design, safety, reliability or robustness analysis. Event chains and real-time requirements are just an additional perspective.

In [3] the starting point for the different analysis is an AMALTHEA model. Timing models with a focus on event chains are automatically generated for INCHRON's Tool-Suite model on different architecture levels from the AMALTHEA. The simulator chronSIM visualizes the event chains and evaluates the real-time requirements.

An example for the visualization of a safety and real-time critical event chain (yellow arrows) for an emergency brake assist is shown in Figure 2. The rows are showing the operating system states of the tasks. If it is blue, the task is running. If it is green, the task is preempted by an ISR or higher priority task. The data of a radar sensor is read and pre-processed on an Infineon AURIX microcontroller. The communication with the high performance computer (Renesas R-Car) is done via Ethernet. The latency of the event chain is composed of execution times of the software, preemption of tasks and communication latencies.

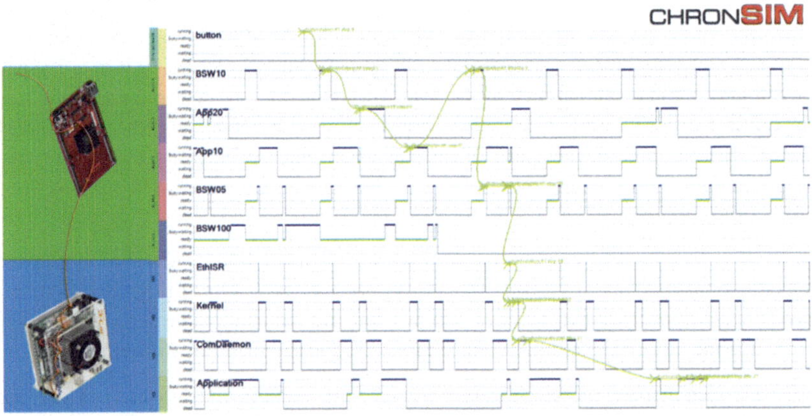

Figure 2: Event chain of an emergency brake assist on software architecture level

Optimization of the desired timing behavior using the model-based approach is significantly faster and cheaper than trying to improve it after the implementation has started. Due to the complexity, high quality constraints and short time-to-market of autonomous driving system, it is no longer possible to fix architecture problems in a later stage of the development process.

5 Architecture Levels

In order to handle complexity, systems are designed on different architecture levels. Starting with a high-level view, more and more details are added at each level. Typical architecture levels are shown in Figure 3. On each architecture level event chains are identified and specified. An event chain step is a reference to a modeling element of the architecture, but it is not a part of the architecture itself. This is a small but important difference. The goals for each architecture level are:

- Common understanding of the data flow between all stakeholders / suppliers to simplify the cooperation and make it more efficient.
- Precise specification of explicit and implicit real-time requirements for the current and following level to allow the architect to design a robust and scalable architecture.
- Verified real-time requirements to avoid sporadic errors, cost and time-consuming re-designs and to ensure the quality of the system.
- Negotiated time budgets for each execution block and communication latencies requirements for the implementation. This avoids time-consuming re-negotiations in the event of an error in a later stage.

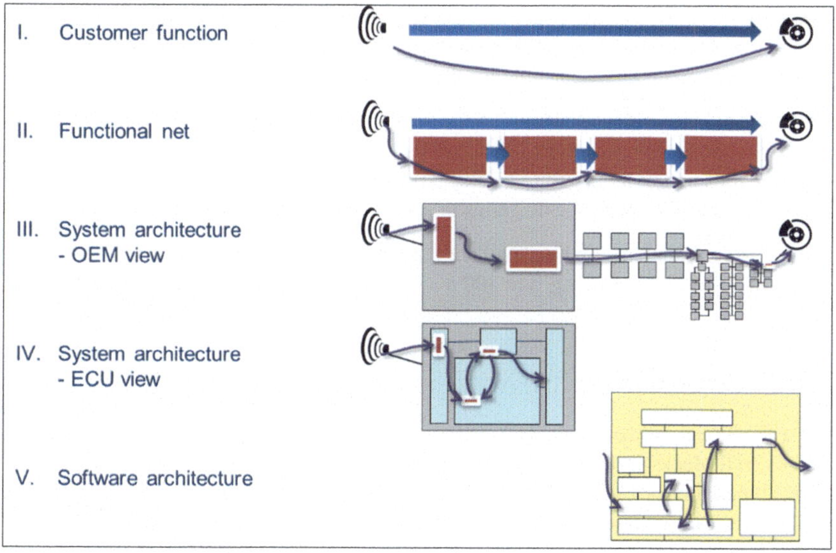

Figure 3: Architecture levels

In order to achieve this goal, on each architecture level the steps are similar:

- Refinement and specification of the event chain.
- Derivation and specification of extrinsic and intrinsic requirements. Additional requirements are often added.
- Negotiation of the timing budgets.
- Evaluation of the architecture.
- Optimization of the architecture, if necessary.
- Derivation of requirements for the next architecture level.

This is an incremental and iterative process with each architecture level and between architecture levels. Many customer projects have shown that the necessary efforts are worthwhile. In [3] the advantages are shown for driver assistant system: "We have found errors in simulation 12 months earlier, and we were able to understand the root cause and to fix it efficiently."

Here is an overview of the architecture levels:

5.1 Customer function

The customer function level (I) is the starting point of our considerations on different abstraction levels. At many OEMs the role of a customer function owner is already existing. He should be responsible for the event chains in general.

The event chains on this level seems to be very simple. It is just a combination of sensors and the corresponding actuators. Effort needs to be spent to get a link between use cases and scenarios on this architecture level and the functional net. To fill this gap SysML and UML diagrams like sequence diagrams or activity diagrams are very helpful.

Real-time requirements on this level are typically derived from physical properties and from driver expectations, e.g. the 300 ms end-to-end latency for an emergency brake assist.

5.2 Functional net

On a functional network level (II) customer functions are decomposed into functional blocks (red blocks in Figure 2) and connectors (blue arrows in Figure 2). Each functional block processes data along the data flow. An event chain step is a reference to a function block, the interfaces/ports of the function blocks and their signals. The functional blocks itself can have internal behavior as well as intrinsic real-time requirements, e.g. maximum data age, start-to-start jitter or periodicity of a closed loop controller. The functional developers have to specify the intrinsic real-time requirements.

The functional architect splits the end-to-end latency into time-budgets (min and max) for the functional blocks and communication channels. This is a very important step in which the architect has to negotiate the time budgets with the functional developer. Based on the timing budgets of the functional blocks and communication channels, the end-to-end latency requirement and the intrinsic real-time requirements of the functional bocks have to be evaluated.

The role of the timing-budget on the functional architecture level changes into requirements on the system architecture level. That means the system architecture has to fulfill the negotiated timing budgets. In general, verified real-time requirements for the next architecture level are the main goal of the event chain evaluation on the functional net level.

5.3 System architecture – OEM and ECU view

The typical OEM view on the system architecture (III) includes ECU and networks. The functional blocks of an event chain are mapped to ECUs (grey boxes in Figure 2) and the some of the connectors are mapped to busses. An important task is to determine the minimum and maximum communication latencies for the communication between ECUs.

If the timing budgets for a communication on the functional net level is too short for the communication on a vehicle bus, a new timing budget has to be negotiated on the functional net level. This means, changing of requirements on a lower architecture level implies a new evaluation on the more abstract level. Therefore, it is important to establish an incremental, iterative process. The goal of the event chain evaluation on this system level is to derive the event chain latencies for each involved ECUs as an important input to the next architecture level.

A Tier-1 supplier is usually responsible for the ECU system architecture (IV). A single functional block of an event chain and its timing budget will be divided into components. Each component is attributed with its intrinsic real-time requirements and the input and output data size. The next step is the evaluation of the ECU end-to-end latency of the event chain and the intrinsic real-time requirements. At this point the amount of data and the required memory bandwidth should also be taken into account.

5.4 Software architecture

Classically the focus on timing on the software architecture level (V) was the mapping of software components to function/runnables, runnables to tasks, and the operating system scheduler strategy. Processing resources are usually limited because of price, space and power consumption. The overall functionality of a system has to utilize the available resources in an intelligent and efficient way. Designing and optimizing the scheduler has a big impact on the performance and requirement fulfillment. Typical real-time requirements are execution times, response times and slack times. An important task of the functional and software developer is to provide these real-time requirements to the software architect and integrator.

On a multi-core system multiple tasks run in parallel and usually share a part of the memory and memory busses. The mapping of tasks and ISRs to cores are crucial because it can extend the latency of an event chain dramatically and can lead to memory bus overload situations. In the latter case it results in the extension of the execution time of the software due to concurrent access to memory from different cores and the resulting wait states. Therefore, the data size for read and write accesses has to specified in the software architecture model.

The design and optimization of the software architecture of high-performance computer is a complex task. Reliability, safety and robustness are important design constraints. To handle the large number of software components from different suppliers and their real-time requirements an efficient process is needed. To focus on event chains gives a common understanding between different stakeholders on the data flows that are crucial for autonomous driving application. An event chain driven architectural design is a proven approach to handle the complexity.

In [2] a process with a focus on event chains on several architecture levels for a safety critical electrical power steering ECU is shown in detail, in [3] for an advanced driver assistant system. In both cases studies a key issue for success was the optimization of event chains.

6 Continuous integration, virtual verification and testing of event chains

Nowadays, most car OEMs are doing black-box testing of ECUs by examining the communication behavior on the connecting bus networks. Usually, the dynamic end-to-end behavior of event chains is only evaluated during system testing, which is no longer sufficient given the complexity of autonomous driving systems. Verifying the quality and reliability of event chains for autonomous driving is a cross-cutting concern where OEMs need to be more involved in integrating and testing individual ECUs and subsystems. [2]

In addition to avoid billions of test kilometres on the road virtual verification methods are needed. Using a continuous integration process an evaluation of the event chains on the different abstraction levels should also be done continuously. In each nightly/weekly build or at lease after a sprint the event chain requirements have to be evaluated. This saves considerable costs, because the errors are found when they arise.

Very often real-time errors occur after the entire software is implemented. For example, a longer execution time causes a response time violation of a low priority task. This butterfly effect is for event chains even more critical to estimate.

On the target hardware the test scope is always the already implemented software. In model-based approach the implemented software in combination with the not yet implemented software can be verified virtually. This means the scope of the virtual verification is always the whole system and not only parts of it.

The specification of the event chains and the real-time requirements are a basic input for a test specification. For each event chain step a measurement point in the software, hardware or on the vehicle bus will be identified. During a test on an evaluation board, HIL or vehicle the event chain steps and their point in time of occurrence are measured. Based on such a measurement trace the event chain requirements can be evaluated continuously or post mortem. For an automatically evaluation our trace visualization and analysis tool chronVIEW is already in place at many customers.

7 Event chains as a communication tool

So, who should take responsibility for event chains as part of the overall system functionality? In general, the stakeholders that are responsible for a customer function must also define the requirements for event chains and other cross-cutting properties, which may have direct influence on the respective function. Many companies have already recognized that it is necessary to strengthen the awareness and knowledge of event chains at several architecture level, and have started to establish dedicated master roles that take responsibility exclusively for event chains. These experts are authorized to define (timing and performance) requirements at different architecture level and take the necessary measures to enforce compliance by all stakeholders. An important task is not only to pay attention to the communication between ECUs, but also to the timing within individual ECUs, in order to anticipate failure propagation and to identify optimization potentials. [2]

8 Summary

In autonomous driving applications it is expected to have up to 1000 safety and real-time critical event chains. Many customer project have shown that a manually consideration of event chains is error prone and time- and cost-consuming. In this article a model-based methodology is presented that focusses an event chains during the whole development process. The basic idea is to evaluate event chain on different architecture level in order to design robust and scalable architectures.

The approach presented is already establish at OEMs and supplier. It avoids sporadic errors in a late stage of the development process.

8.1 Authors

Olaf Schmidt
Business Development Manager
olaf.schmidt@inchron.com

Dr. Ralf Münzenberger
CEO and co-founder
ralf.muenzenberger@inchron.com

8.2 About INCHRON GmbH

INCHRON is the world-wide leading provider of solutions for architecture, design, and automated optimization of real-time systems with regards to timing behavior and performance. Our solutions cover the whole range from single-core to multi-core to multi-CPU to distributed systems. Well-known OEMs and tier suppliers world-wide rely on INCHRON's methods and tools. Our solutions are used successfully in industries such as automotive, automation, avionics, defense, healthcare, mobile and M2M/IoT, covering all phases of the development lifecycle. Since INCHRON was founded in 2003, our consulting team has made significant contributions to more than 170 successful customer projects.

The INCHRON Tool-Suite provides powerful tools, covering simulation, worst-case analysis, automated optimization, and comprehensive visualization and analysis of traces. The Tool-Suite covers all important steps to focus on event chains.

INCHRON GmbH, Karl-Liebknecht-Str. 138, 14482 Potsdam, Germany
Phone +49-331-2797892-0, E-Mail info@inchron.com, Web www.inchron.com

8.3 References

[1] Evaluation of Architecture Variants for Hard Real-Time Systems, Dr.-Ing. Isabella Stilkerich, Schaeffler Technologies AG & Co. KG; Frank Sigiliano Pinecker, Methodpark Consulting GmbH; ISO 26262 9th International Annual Conference 2017

[2] Mesut Özhan, Olaf Schmidt, Chain Event Analysis for Autonomous Driving, Whitepaper, INCHRON GmbH

[3] A Practical Approach to the Simulation of Safety-critical Automotive Control Systems considering Complex Data Flows, Sébastien Dubé, Hella Engineering France, Mesut Özhan, INCHRON GmbH, Achim Rettberg, Hella Electronics, proceedings of Embedded Real Time Software and Systems Congress, January 2016

[4] T. Jäger, I. Houben and R. Münzenberger, "Model based architecture development and simulation," in *ESE Kongress*, Sindelfingen, 2015.

[5] https://www.inchron.com/automotive/

High-performance data acquisition and replay

Thomas Schöpfner, ETAS GmbH

T. Bertram (Hrsg.), *Automatisiertes Fahren 2019*, Proceedings,
https://doi.org/10.1007/978-3-658-27990-5_5

DRIVING EMBEDDED EXCELLENCE
ATZ: ATZ: Automatisiertes Fahren 2019
High performance data acquisition and replay
Wiesbaden, Apr.2019 Thomas Schöpfner

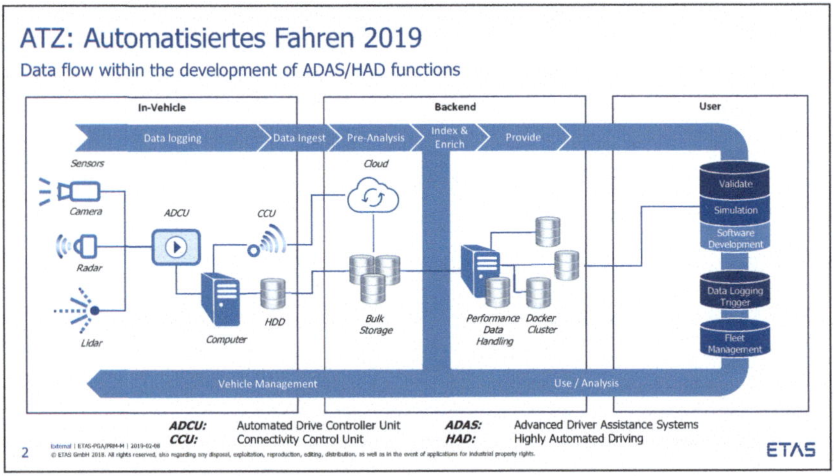
ATZ: Automatisiertes Fahren 2019
Data flow within the development of ADAS/HAD functions
In-Vehicle
Backend
User
Data logging
Data Ingest
Pre-Analysis
Index & Enrich
Provide
Sensors
Camera
Radar
Lidar
ADCU
CCU
Computer
HDD
Cloud
Bulk Storage
Performance Data Handling
Docker Cluster
Validate
Simulation
Software Development
Data Logging Trigger
Fleet Management
Vehicle Management
Use / Analysis
ADCU: Automated Drive Controller Unit
CCU: Connectivity Control Unit
ADAS: Advanced Driver Assistance Systems
HAD: Highly Automated Driving
2
External | ETAS-PGA/PRM-M | 2019-02-08
© ETAS GmbH 2018. All rights reserved, also regarding any disposal, exploitation, reproduction, editing, distribution, as well as in the event of applications for industrial property rights.

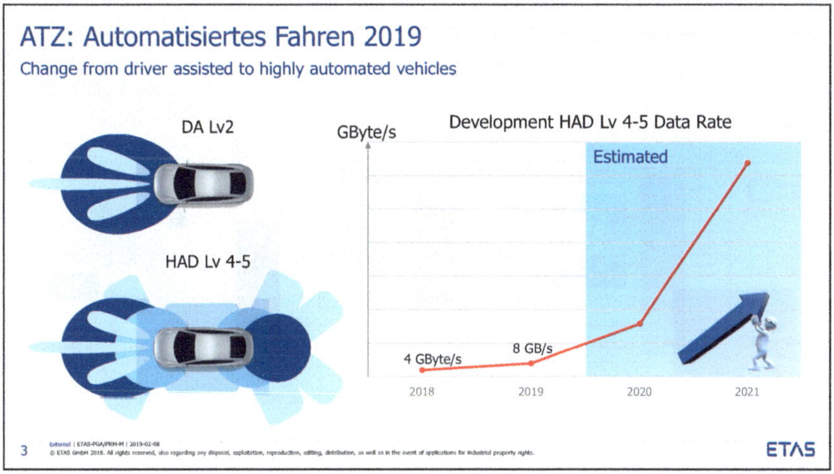
ATZ: Automatisiertes Fahren 2019
Change from driver assisted to highly automated vehicles
DA Lv2
HAD Lv 4-5
GByte/s
Development HAD Lv 4-5 Data Rate
Estimated
4 GByte/s
8 GB/s
2018
2019
2020
2021
3
External | ETAS-PGA/PRM-M | 2019-02-08
© ETAS GmbH 2018. All rights reserved, also regarding any disposal, exploitation, reproduction, editing, distribution, as well as in the event of applications for industrial property rights.
ETAS

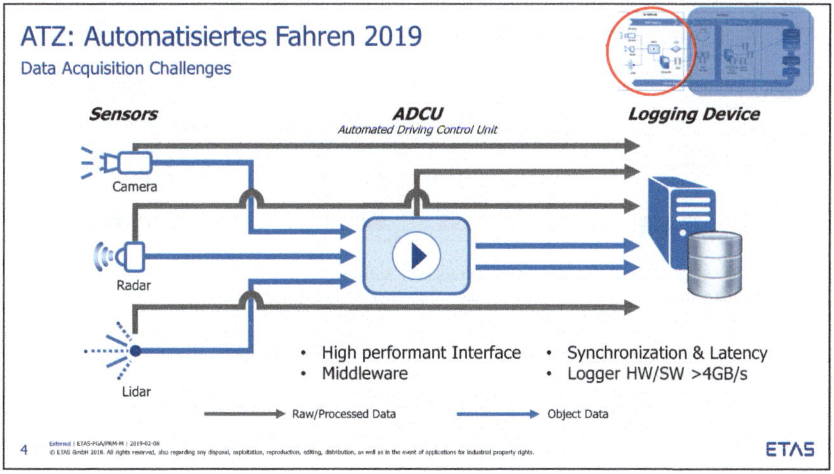
ATZ: Automatisiertes Fahren 2019
Data Acquisition Challenges
Sensors
ADCU
Automated Driving Control Unit
Logging Device
Camera
Radar
Lidar
• High performant Interface
• Middleware
• Synchronization & Latency
• Logger HW/SW >4GB/s
Raw/Processed Data
Object Data
4
External | ETAS-PGA/PRM-M | 2019-02-08
© ETAS GmbH 2018. All rights reserved, also regarding any disposal, exploitation, reproduction, editing, distribution, as well as in the event of applications for industrial property rights.
ETAS

ATZ: Automatisiertes Fahren 2019

ADCU µC/µP: Data Acquisition System with GETK-P

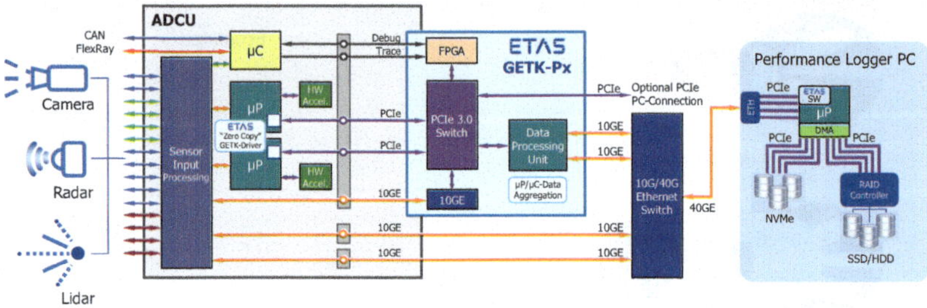

5 ETAS

ATZ: Automatisiertes Fahren 2019

GETK Software Integration: In-Vehicle SoC with GETK

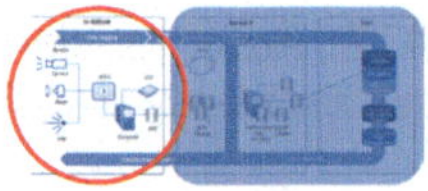

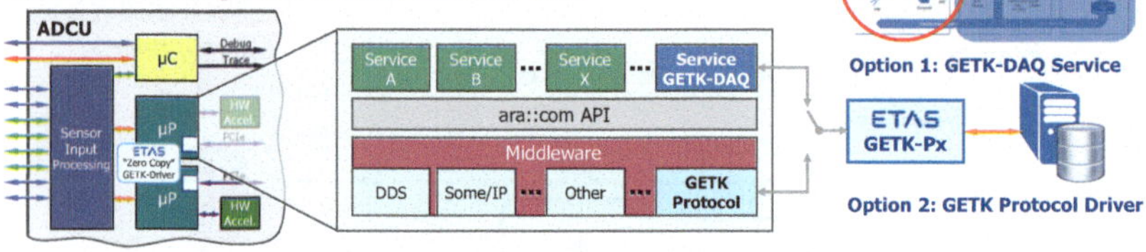

2 Options for GETK communication with µProcessor Software

1. **GETK-DAQ Service:** GETK communication as a **Data Acquisition Service** above ara::com API
   - \+ General approach based on standardized ara::com API: Independent of actual Middleware
   - o DAQ system load depends on Middleware implementation e.g. support of "Zero Copy" transport methods
   - – High dependency on the monitored services: Service interface modifications require GETK-DAQ Service adaptation
2. **GETK Protocol Driver:** GETK communication driver integrated into Middleware
   - \+ Independent of the monitored services: Service interface modifications have no impact on the GETK Protocol Driver
   - o DAQ system load depends on GETK Protocol Driver implementation including "Zero Copy" transport methods
   - – Specific approach, highly dependent on the actual Middleware

6 ETAS

ATZ: Automatisiertes Fahren 2019

Simulation Use Cases

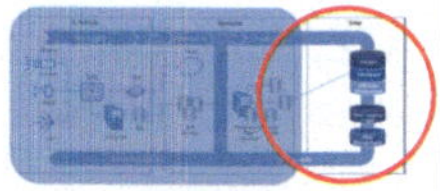

Development Support

- Faster development
- Early prototyping

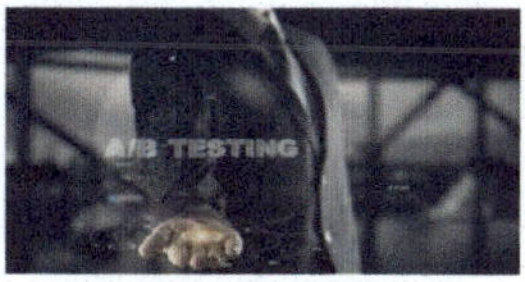

Verification

- Test against specifications

Validation

- Test for unintended behaviors and violate assumption

7 External | ETAS-PGA/PRM-M | 2019-02-08

ETAS

ATZ: Automatisiertes Fahren 2019

Technical variant: Entire User System Replay

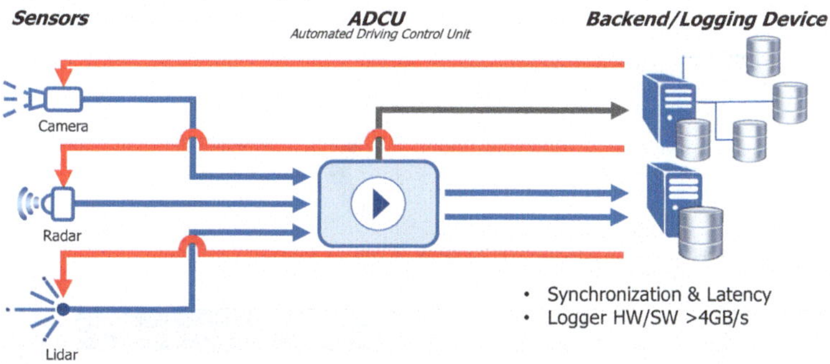

- Synchronization & Latency
- Logger HW/SW >4GB/s

Raw/Processed Data from Data Storage — Raw/Processed Data — Object Data

8 External | ETAS-PGA/PRM-M | 2019-02-08

ETAS

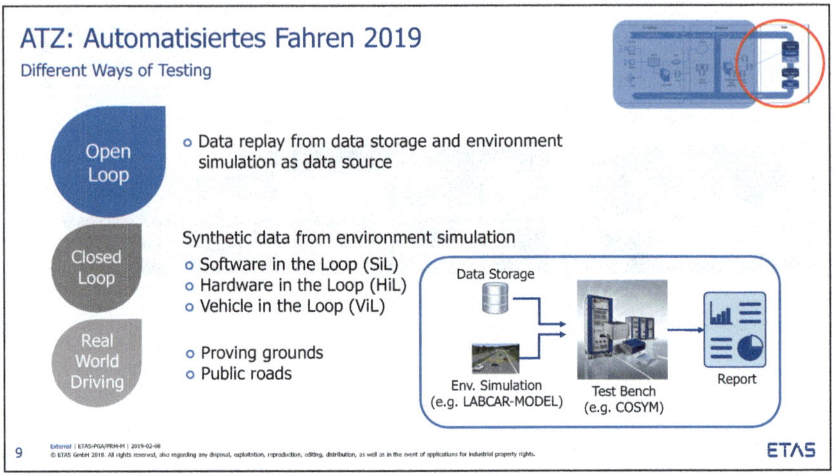
ATZ: Automatisiertes Fahren 2019
Different Ways of Testing
Open Loop
Data replay from data storage and environment simulation as data source
Closed Loop
Synthetic data from environment simulation
Software in the Loop (SiL)
Hardware in the Loop (HiL)
Vehicle in the Loop (ViL)
Real World Driving
Proving grounds
Public roads
Data Storage
Env. Simulation (e.g. LABCAR-MODEL)
Test Bench (e.g. COSYM)
Report
9
ETAS

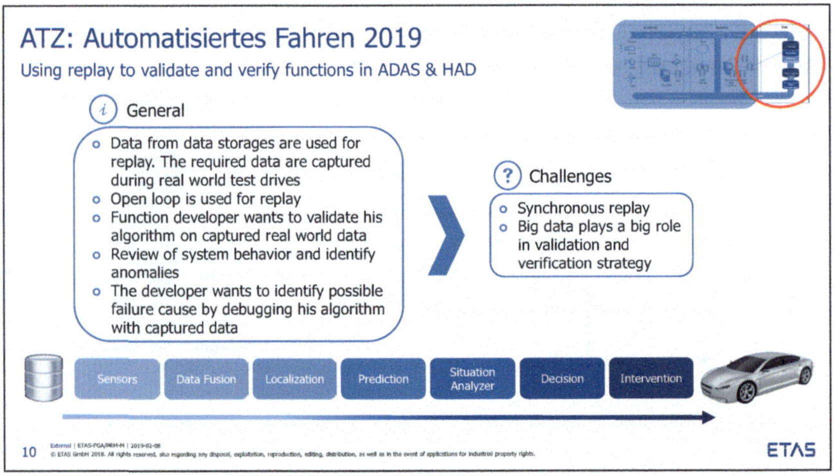
ATZ: Automatisiertes Fahren 2019
Using replay to validate and verify functions in ADAS & HAD
General
Data from data storages are used for replay. The required data are captured during real world test drives
Open loop is used for replay
Function developer wants to validate his algorithm on captured real world data
Review of system behavior and identify anomalies
The developer wants to identify possible failure cause by debugging his algorithm with captured data
Challenges
Synchronous replay
Big data plays a big role in validation and verification strategy
Sensors
Data Fusion
Localization
Prediction
Situation Analyzer
Decision
Intervention
10
ETAS

ATZ: Automatisiertes Fahren 2019

Summary and Outlook

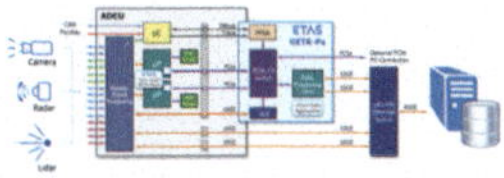

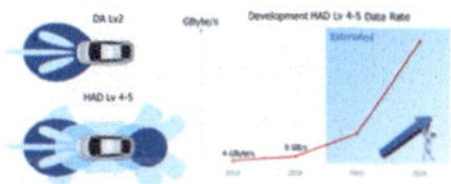

Summary

- PCIe Interface will be the standard interface in data acquisition
- Ethernet will be the standard for the in-vehicle network architecture
- Data amount will increase due to amount and resolution of sensors

Outlook

- Data amount will increase with the development of highly automated driving projects →Big Data
- Simulation during development will play a dominant role in the future
- High performance PC, Virtual control units, sensors and GETK will be the first step in the development while sensors and/or control units are still in development

ETAS

DRIVING EMBEDDED EXCELLENCE

Thank you

Driver assistance systems and automated driving functions – impact potentials, challenges and solutions from the point of view of the AZT

Dr. Johann Gwehenberger, Dr. Christoph Lauterwasser,
Marcel Borrack, Melanie Kreutner, Carsten Reinkemeyer

AZT Automotive GmbH – Allianz Zentrum für Technik

T. Bertram (Hrsg.), *Automatisiertes Fahren 2019*, Proceedings,
https://doi.org/10.1007/978-3-658-27990-5_6

1 Introduction

It has been demonstrated that driver assistance systems are making a significant contribution toward increasing road safety. For more than a decade, the Accident Research Department of the Allianz Center for Technology (AZT) has been analyzing impact potentials as part of several research projects, including AKTIV, TRACE and euroFOT [1, 2, 3], and was largely responsible for the fact that effective driver assistance systems are now taken into account in several countries when pricing insurance premiums. Potential effectiveness is regularly assessed on the basis of third-party liability (TPL) and motor own damage (MoD) claims reported to Allianz Versicherung [4, 5, 6, 7]. In this context, AZT – in cooperation with automobile manufacturers Audi, Daimler and Volvo – has investigated the efficiency and the limits of front collision warning systems and autonomous emergency braking systems in passenger cars. The results of these studies are discussed in Chapter 2. Another focus of the paper is a pilot study (chapter 3) on forecasting the influence of automated driving functions on the claims trend in Germany, which is being continued as part of the EU research project L3Pilot.

For highly and fully automated driving to be accepted by society, accidents involving automated vehicles must be clarifiable with regard to liability and responsibility. It must also be possible to monitor and evaluate the safety of automated systems. Against this background, the AHEAD working group has developed a data model for accident investigation in vehicles with highly automated driving functions of SAE automation level 3 and above (see chapter 4).

2 Ex-post Analyses of the Effectiveness of AEB

Autonomous emergency braking systems (AEB) and front collision warning systems are supposed to help avoiding accidents in longitudinal traffic. Numerous studies of AZT accident databases with property damage and personal injury show a high relevance (= maximum theoretical accident avoidance potential) for AEB in longitudinal traffic (13 to 36%). AEB systems that react to pedestrians and cyclists are particularly relevant in accidents involving personal injury (18%) [5, 6].

Furthermore, the AZT has studied the actual influence on rear-end collisions of various generations of these driver assistance systems from the manufacturers Audi, Mercedes-Benz and Volvo, as well as those in electric and plug-in hybrid vehicles.

First, the claims in each aggregated claims portfolio were transferred into a four-field matrix (see Table 1). The main precondition here is knowing whether the respective colliding vehicle was equipped with an appropriate ADAS or not (e.g., information provided by the manufacturer or a specific Audatex query).

Table 1: Four-field matrix for calculating the relative risk (RR)

	AEB-relevant rear-end collision	*not AEB-relevant no rear-end collision*
with AEB	***a***	***b***
without AEB	***c***	***d***

These data can then be used in Formula 1 to calculate the relative risk (RR). The relative risk shows the ratio of probabilities that an accident will occur with or without AEB, with the relevant accidents compared to all accidents in the respective category.

$$\boldsymbol{RR} = \frac{\frac{\boldsymbol{AEB\ relevant\ accidents_{Vehicle\ with\ AEB}}}{\boldsymbol{all\ accidents_{Vehicle\ with\ AEB}}}}{\frac{\boldsymbol{AEB\ relevant\ accidents_{Vehicle\ without\ AEB}}}{\boldsymbol{all\ accidents_{Vehicle\ without\ AEB}}}} \left(= \frac{\frac{a}{a+b}}{\frac{c}{c+d}}\right) \tag{1}$$

The related benefit (ADAS) can then be calculated by subtracting the relative risk from 1 and multiplied by 100 to show the percentage (Formula 2). Ultimately, the benefit is a value that indicates how much less likely a vehicle equipped with AEB is to cause a rear-end collision than is a vehicle without AEB.

$$Benefit_{AEB} = (1 - RR) * 100\% \tag{2}$$

Figure 1 summarizes the results of current studies using comprehensive collision claims as an example.

The first positive aspect that needs to be emphasized is that in each of the individual studies, the assistance systems provided a benefit of at least 10% in terms of reducing the frequency of rear-end collisions. However, by definition the study results are subject to multivariate influences and therefore are dependent on many factors. These primarily include the sample size, the generation of the driver assistance system, different driver clientele and driver behavior, and, in particular, limitations on choices of comparable vehicle group. In addition, in reviewing the individual studies, because the number of cases is sometimes not big enough, the only significant results are the results of the analysis of the Mercedes B-Class (W246) "with Collision Prevention Assist (CPA)" with a 95% confidence interval and the overall review of all studies conducted.

As for the respective comparison group, it should be emphasized that for the vehicle models Audi A8 (D4) and Mercedes E-Class (W212), it was possible to compare the claims within the same model in each case. For these vehicles, a front collision avoidance system was available as an option, as a result, it was possible to form sufficiently large clusters based on the extent to which the respective vehicles were equipped ("with AEB" and "without AEB").

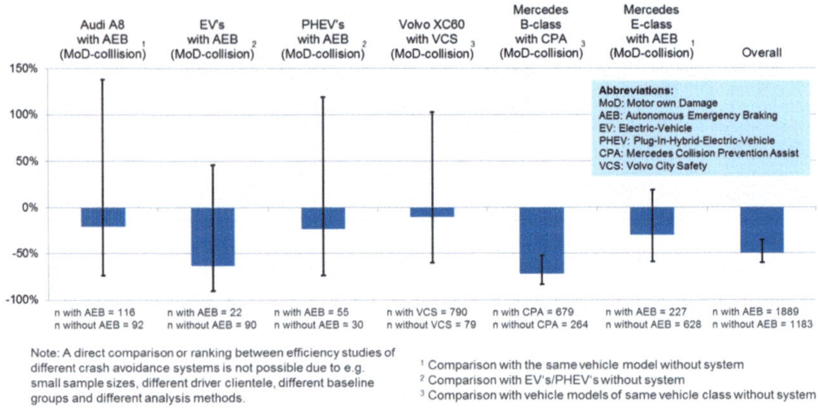

Figure 1: Overview of study results regarding reductions in frequency of rear-end collisions by various rear-end collision avoidance systems [8, 9, 10]

To analyze electric vehicles (EVs) and plug-in hybrid electric vehicles (PHEVs), the vehicles were grouped as per drive technology, because a manufacturer-based or model-based comparison was not possible as the number of cases was too small. Nevertheless, this approach made it possible to form sufficiently large numbers of cases and it was at least possible to compare the frequencies of rear-end collisions of EVs/PHEVs with AEB and of EVs/PHEVs without AEB. The claims included a total of 15 car brands, among them models of the manufacturers BMW, Mitsubishi, Renault, Toyota, VW, Mercedes-Benz and Tesla.

In contrast, when the Volvo XC60 was introduced in 2008, the Volvo City Safety (VCS) autonomous emergency braking system was already standard equipment, which is why comprehensive collisions of passenger cars in the same SUV vehicle category from the 2011 in-depth database were used as the comparison group. The calculated VCS benefit, which was obviously limited compared to the other analyses shown (reduction around 10%) can be explained by the initially lower performance of the VCS based on LIDAR sensors. For example, the first generation of the system was initially active only up to a speed of 30 km/h and was intended to prevent accidents up to a differential speed of 15 km/h [11]. Moreover, the system did not yet include any front collision warning system, the VCS activated automatic braking only in the event of an imminent collision.

It was especially hard to select a suitable vehicle for comparison with the Mercedes B-Class (W246). This model belongs to the compact class, but is primarily driven by older drivers (55+) whose driving profile (e.g., place, time, distances, mileage/driving performance and number of parking and maneuvering steps) is significantly different from drivers of other compact class vehicle models. There is a high probability that these factors are responsible for the fact that the Mercedes Collision Prevention Assist yields a particularly high benefit – a 70% lower probability of rear-end collisions – compared to comparable vehicles without any system.

Conclusion of the ex-post analyses:

Despite the aforementioned limitations on the ex-post analysis of accidents in vehicles with or without AEB, one can conclude that front collision avoidance systems help to reduce the frequency of rear-end collisions. A review of all individual studies with a sufficiently large number of cases (n with AEB = 1,889, n without AEB = 1,183) yields a benefit of nearly 50% (see Figure 3, the last bar on the right-hand side of the chart). Further research results on autonomous emergency braking systems compiled by the AZT and analyses published by the Insurance Institute for Highway Safety in the USA (IIHS/HLDI) confirm that the level of effectiveness is between 40% and 50% [12, 13].

Conversely, however, that also means that there is further potential for improvement in the remaining 50% to 60% of rear-end collisions (e.g., detecting cut-in situations, reacting appropriately when the vehicle overlap is small).

From the standpoint of the consumer and the insurance industry, the trend toward higher repair costs will also play an important role if an accident occurs that results in property damage despite the available assistance systems. Suitable steps should be taken to counter this trend (e.g., self-adjustment of sensors, ease of repair of sensors).

3 Accident Avoidance Potential of L3+Functions

The objective of the European research project L3Pilot is to test 100 vehicles with SAE level 3 and 4 functions under real traffic conditions. One of the tasks of the AZT in this project is to determine the expected safety benefit through the following four automated driving functions based on insurance claims: Motorway Pilot, Traffic-Jam Pilot, Urban Pilot und Parking Pilot (see figure 2).

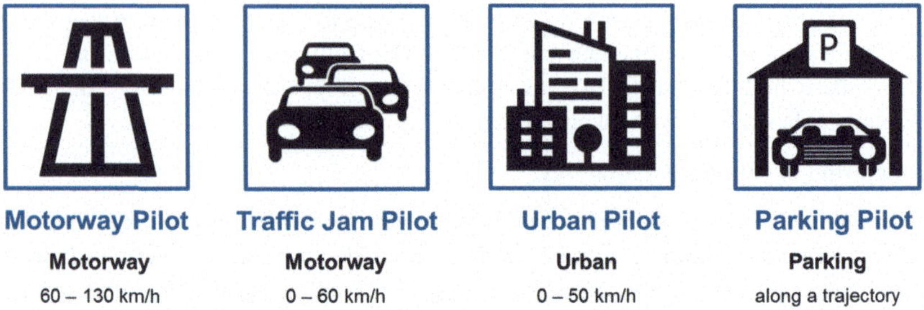

Figure 2: Automated driving functions at SAE levels 3 and 4 defined in the L3Pilot project [14, 15]

As part of Isabella Ostermaier's master thesis the AZT claims databases were used to determine the expected safety benefit [15]. The 15,600 claims included form a representative random sample of the motor third party liability and motor own damage claims reported to Allianz. Based on the accidents contained in the claims databases, the number of accidents that can be avoided by the L3+ functions is influenced by the following four key figures (see figure 3).

- **Relevance**: The theoretical maximum percentage of claims that can be addressed by the L3+functions.
- **Market penetration**: The percentage of cars on the road that are equipped with L3+functions.
- **Efficiency**: The percentage of relevance that can be achieved under real road traffic conditions.
- **Utilization**: The extent to which the technical system is used by drivers. [16]

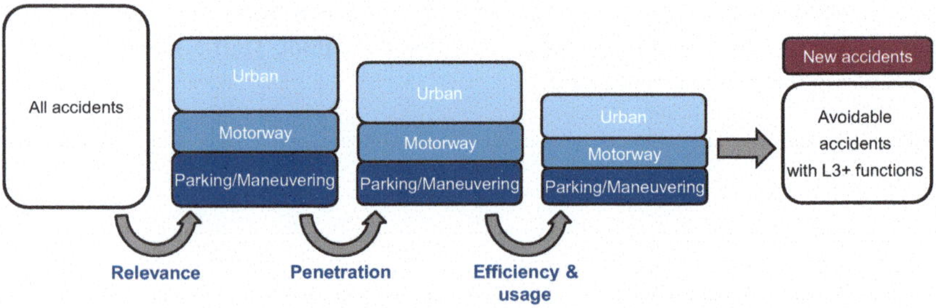

Figure 3: Method to evaluate the expected safety benefit of L3+ driving functions (schematic) [15]

Multiplying the four ratios for the respective automated driving function will yield the potential for reducing claims for the period under review after the market launch. Using the example of the Urban Pilot (see figure 4), a safety benefit of 2.1% for motor own damage collisions can be expected in 20 years after the market launch of the system.

Whereas almost 20 % of all claims considered are theoretically within the scope of the Urban Pilot, this share is reduced to only 2.1 % taking into account market penetration, efficiency and utilization.

Relevance	x	Mean market penetration	x	Efficiency	x	Utilization	=	Expected benefit
19.1 %	**x**	**61.0 %**	**x**	**37.6 %**	**x**	**48.0 %**	**=**	**2.1 %**

Figure 4: Expected safety benefit by the Urban Pilot 20 years after market launch of the system

Figure 5 shows an example of the forecast for the individual L3+ functions in 20 years after market launch.

In summary, the expected claims reduction is between 5.0% and 6.8%, depending on the insurance segment. One reason for these low percentages is that not every new vehicle will be equipped with the L3+ functions as soon as the automated driving functions become available on the market. In addition, the vehicle fleet will continue to include older vehicles with only a few assistance systems.

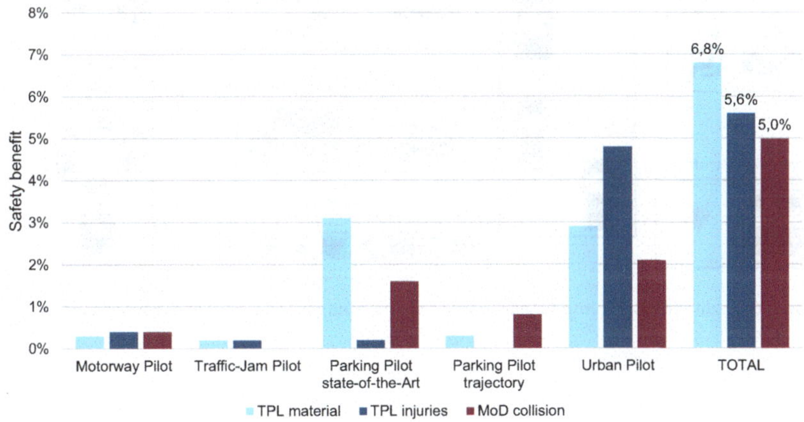

Figure 5: Expected safety benefits of the investigated L3+ functions [15]

In order to be able to represent the future accident occurrence in all its changing aspects, it is necessary to consider not only the positive effects on road safety, but also the new risks resulting from the changed task sharing between man and machine. In order to identify critical driving situations in future traffic, a workshop was held with experts in automated driving. The eight driving scenarios that were determined for evaluation are illustrated in Figure 6.

Takeover situations can increase the risk of accidents by the fact that passenger car drivers cannot perform the driving task in the corresponding time because they turned away from the driving task for a longer period of time. Particularly on German motorway sections without speed limits, critical situations can arise when changing lanes. Furthermore, the lack of communication between road users, especially in merging traffic with automated and non-automated vehicles, increases the risk of accidents. In contrast to the human driver, the automated driving function lacks driving experience and anticipation in order to react situatively to influencing factors such as road conditions and other road users. This can lead to an increase in critical situations in connection with environmental conditions, obstacles, stop and go traffic, the formation of a rescue alley and lateral offset.

In addition to new situations, such as taking over the driving task, some of the scenarios (e.g. lane change, lateral offset, ...) involve accident types that already exist in traffic, but whose frequency of occurrence will change in future accident events.

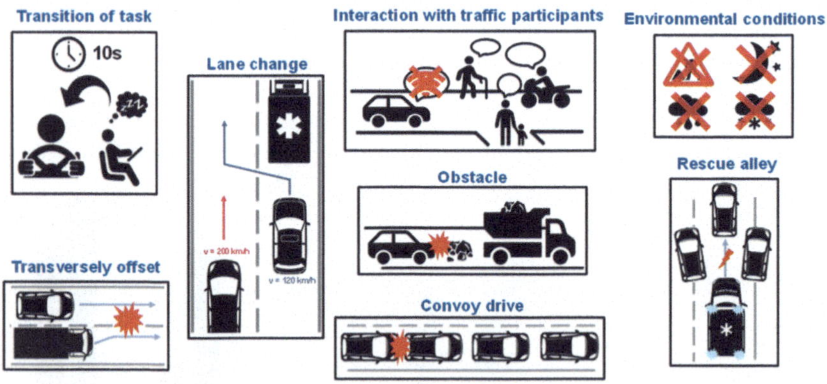

Figure 6: Eight critical driving scenarios due to increasing automation of the driving task [15]

4 Event Data Recorder for Vehicles with L3+Functions

In the event of an accident, modern motor vehicles store a series of event data that are of great and increasing importance for accident clarification. However, even for experts it is not transparent in which vehicle models which data is recorded in which quality and how or whether it can be read. While, for example, a minimum standard for data recording in vehicles has existed for years in the USA, there is still no corresponding regulation in the EU.

In Germany, the amendment of the Road Traffic Act (§ 63a StVG) for the first time regulated data processing including data recording in the sense of a driving mode storage device (DSSAD) for highly and fully automated vehicles. However, the data elements defined therein are not sufficient to clarify the causes of accidents and the corresponding liability issues. The regulation does not specify accident recognition by the system, data security and access options. In order for highly and fully automated driving to be widely accepted by the society, it must be possible to analyse accidents involving these vehicles with regard to liability and responsibility. At the same time, victim protection must be guaranteed and it must be possible to monitor and evaluate the safety of automated systems. Accident data must therefore be treated separately from commercially usable vehicle data.

Against this background, the **AHEAD** working group (**A**ggregated **H**omologation-proposal for **E**ventRecorderData for **A**utomated **D**riving), consisting of the core partners CARISSMA, AXA, DEKRA, Continental and Allianz, has developed a data model for accident investigation in vehicles with highly automated functions with SAE automation level 3 and above [17]. The data recording shall be limited to the time period necessary for the clarification. A continuous storage of general driving data is not required. Data privacy and customer protection are top priorities when defining the data model. The customer shall have the decision whether data beyond the mandatory minimum will be recorded.

The proposed standardisation comprises a catalogue of necessary data elements, trigger thresholds for storage and possibilities for data processing and initially refers to motor vehicles of EC classes M1, M1G, M2, N1, N2 and N3. The data elements to be stored are divided into 4 standardised categories (see Figure 7).

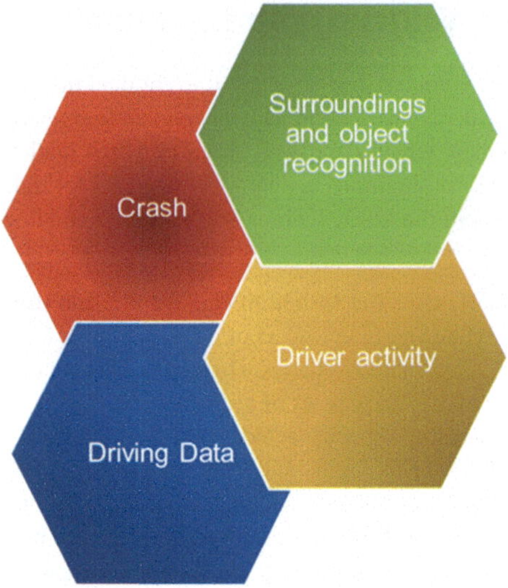

Figure 7: Main data categories of the AHEAD data model (schematic) [17]

The AHEAD data model includes but is not limited to the following data:

Driving data

- Vehicle status, operating mode (e.g. manual, automated, remote-controlled), speed, yaw angle, control interventions of the assistance system, takeover request
- Diagnostic data of safety-relevant systems and components (status, system failures/technical malfunctions)…

Driver activity

- Video feeds of interior cameras, steering, seating position, pedal positions, driver activity…

Environment and object recognition

- Video feeds from front and rear view cameras, sensor data, classified objects, object position, object direction, object velocity, calculated motion…

Crash

- Date, timestamp, location, acceleration, collision speed, seat belt status, airbag, restraint system
- Trigger sensor data

The trigger threshold for data storage must be defined by an advanced algorithm in such a way that even accidents with low accelerations and low speed changes, e.g. accidents with vulnerable road users, reliably lead to storage. The given US requirement of $\Delta v = 8$ km/h is not applicable for e.g. maneuvering issues and collisions with pedestrians.

In addition to standardising data elements and trigger thresholds for storage, access to vehicle data must also be regulated. The AHEAD guidelines for non-discriminatory access to vehicle data are:

- Legitimate interest
- Fair and undistorted competition
- Data privacy and data security
- Tamper-proof access and liability
- Data economics
- Standardised interface
- Crash resistance of the data storage system in the vehicle
- Event data storage for a limited period before and after an event (~ 30 sec)

Once the data has been stored in the vehicle, access to it must be guaranteed for authorised persons. Data processing via an independent data trustee would mean fair data access for all authorised persons. Access would be technically simpler and more affordable while respecting privacy. The data trustee can also meet the various requirements for data storage and deletion, manage the data, verify access rights and protect against manipulation.

Compliance with the standards outlined should in future be a prerequisite for the homologation of vehicles with highly automated systems in the European Union.

5 Conclusions and outlook

The ex-ante analyses conducted by the AZT show the high accident avoidance potential of AEB systems. In addition, the presented ex-post analyses of the investigated driver assistance systems show a significant reduction in rear-end collisions (-50%) for the first time. Therefore, any and all measures taken by industry, trade, fleet operators and other organizations that will promote the spread of autonomous emergency braking systems are welcome. As a result, Allianz also supports the plans of the European Commission to require autonomous emergency braking systems for new vehicles in the European Union in the future as part of the General Safety Regulation.

As penetration rates of driver assistance systems increase and as different system designs proliferate, in the future it will also be important to be able to evaluate systems that are not yet represented in the market. In order to be in a position to adequately assess the effectiveness of ADAS with increasing degree of automation, AZT and its partners in the automotive industry and the scientific community will continue to further develop prospective and retrospective methods. An important platform for the further development and harmonization of these methods for safety assessment is the European initiative P.E.A.R.S. (Prospective Effectiveness Assessment for Road Safety) [18, 19].

For the user acceptance and a holistic assessment of the future accident occurrence, the positive and negative effects of automated driving must therefore be determined. The AHEAD data model for accident clarification for vehicles with highly automated driving functions from SAE automation level 3 on is an important key element for this.

Acknowlegments

The authors would especially like to thank Ms. Isabella Ostermaier, Mr. Stefan Sanktjohanser, Mr. Julian Schatz, Mr. Peter Mayrhofer, Mr. Tobias Frank and Mr. Philipp Krebs whose scientific work at the AZT provided an important foundation for estimating the potential and evaluating the efficiency of driver assistance systems. We would also like to thank the partners of AHEAD: AXA, CARRISSMA, Continental and DEKRA for the joint development of the AHEAD position.

Bibliography

[1]: Pfaffenbauer, T.; Gwehenberger, J.; Schwarz S.; Wermuth, G.: Unfallstruktur- und Wirkpotentialanalysen zu den AKTIV-Applikationen auf der Basis von Lkw-Haftpflichtschäden (2010), Allianz Deutschland AG, ISBN 978-3-942022-03.

[2]: Daschner, D.; Gwehenberger, J.; Schwarz, S.; Wermuth, G.; Schönfelder, M.; Hofmann, F.: Unfallstruktur- und Wirkpotentialanalysen zu den AKTIV-Applikationen auf der Basis von Pkw-Haftpflichtschäden mit Personenschaden (2010), Allianz Deutschland AG, ISBN 978-3-942022-02-6.

[3]: Geißler, T.; Bühne, J.; Dobberstein, J.; Gwehenberger, J.; Knieling, M.: Overall Cost-Benefit Study (2012), Research Report D6.7 relating EU Project euroFOT under 7th Framework programme ICT-2-6.2.

[4]: Gwehenberger, J.; Borrack, M.: Influence of Driver Assistance Systems on Insurance Claims, In: ATZ – Automobiltechnische Zeitschrift, p. 60-65, October 2015.

[5]: Gwehenberger, J.; Borrack, M.: Einfluss von Fahrerassistenzsystemen auf Versicherungsschäden. In: VKU – Verkehrsunfall und Fahrzeugtechnik, p. 342-350, October 2015.

[6]: Borrack, M.; Gwehenberger, J.; Schatz, J.; Feig, P.: Aktuelle Forschungsergebnisse zur Wirksamkeit von Fahrerassistenzsystemen mit zunehmendem Automatisierungsgrad. In: VKU – Verkehrsunfall und Fahrzeugtechnik, p. 378-385, November 2017.

[7]: Gwehenberger, J.; Behl, T.; Lauterwasser, C.: Wie wirksam sind Fahrerassistenzsysteme – vom Bagatellschaden bis zum schweren Unfall?. In: VKU – Verkehrsunfall und Fahrzeugtechnik, p. 60-65, February 2012.

[8]: Sanktjohanser, S.: „Aufbau und Analyse einer Vollkasko-Schadendatenbank mit Fahrzeugen der Mercedes-Benz E-Klasse und B-Klasse (W212 und W246)“ (2015), unpublished Bachelor's thesis

[9]: Frank, T.: „Aufbau und Analyse einer Kraftfahrzeughaftpflicht- und Vollkasko-Schadendatenbank mit Fahrzeugen der Baureihe Volvo XC 60“ (2017), unpublished Master's thesis.

[10]: Mayrhofer, P.: „Analyse von Unfällen mit Elektrofahrzeugen zur Ableitung von Maßnahmen der Schadenverhütung und -minderung an Fahrzeugstruktur und fahrzeugspezifischen Bauteilen“ (2018), unpublished Bachelor's thesis.

[11]: Volvo Car Group: "City Safety standard in the new Volvo XC60". 2008. [Online] Available: https://www.media.volvocars.com/global/en-gb/media/pressreleases/14564, accessed on April 16th 2018.

[12]: Fildes, B.; Keall, M.; Bos, N.; Lie, A.; Page, Y.; Pastor, C.; Pennisi, L.; Rizzi, M.; Thomas, P.; Tingvall, C.: Effectiveness of low speed autonomous emergency braking in real-world rear-end crashes. In: Accident Analysis & Prevention 81 (2015), p. 24-29.

[13]: IIHS Status Report newsletter, Vol. 51, No. 1, January 28, 2016.

[14]: EU Projekt L3 Pilot – Driving Automation: https://l3pilot.eu/index.php?id=26,

[15]: Ostermaier, I.: „Analyse von Versicherungsschäden zur Ermittlung des Schadenvermeidungspotentials durch L3+ Funktionen“, TU München (2018), Master's thesis.

[16]: Gesamtverband der deutschen Versicherer: Automatisiertes Fahren – Auswirkungen auf den Schadenaufwand bis 2035 (2017), https://www.gdv.de/resource/blob/8282/c3877649604eaf9ac4483464abf5305d/download-der-studie-data.pdf, accessed on February 12th 2019

[17]: Kreutner, M.; Lauterwasser, C.: Investigating accidents involving highly automated vehicles: concept of data trustee and data model for future homologation. In: Proceedings of EVU Congress 2018, Dubrovnik.

[18]: Page, Y.; Fahrenkrog, F.; Fiorentino, A.; Gwehenberger, J.; Helmer, T.; Lindman, M.; op den Camp, O.; van Rooij, L.; Puch, S.; Fränzle, M.; Sander, U.; Wimmer, P.: A comprehensive and harmonized method for assessing the effectiveness of advanced driver assistance systems by virtual simulation: the P.E.A.R.S. initiative (2015), https://www-esv.nhtsa.dot.gov/Proceedings/24//files/24ESV-000370.PDF, accessed on May 09th 2018.

[19]: https://pearsinitiative.com, accessed on February 10th 2019.

Vehicle Data for Automated Driving over the Vehicles Lifecycle

Dipl.-Ök. Gerald-Alexander Beese

Dipl.-Ing. (FH) Helge Kiebach, MEng (TAR)

T. Bertram (Hrsg.), *Automatisiertes Fahren 2019*, Proceedings,
https://doi.org/10.1007/978-3-658-27990-5_7

1 Introduction

As todays modern state of the art vehicles are underlying incremental changes with regard especially to 'Automated Driving' (e.g. sensor suite, control units, processing units), 'Connectivity' (e.g. software updates, over-the-air, infotainment), 'Electrification' (e.g. battery cells, drive train), and 'Shared Mobility' (e.g. changing customers behaviour, changing markets, different usage of the car), vehicle data gains in importance. Thus, this article focuses on data and data sets generated during the usage of vehicle fleets in general and individual vehicles in particular. In addition to the general relevance of vehicle data for several stakeholders and interested parties, further complexity arises due to the development of vehicles from driver operated mobility devices towards system enhanced automated driving systems (according to the five level SAE approach). With regard to vehicle lifecycles, a holistic consideration of vehicle data lacks of analysis with its consequences for all market participants involved.

2 Definitions and Differentiation

2.1 Vehicle Data

In this essay, vehicle data is being defined as any data generated by the vehicle and / or the vehicles user(s) during the operation of the vehicle, which may either be a physical driver or a system taking over control of the vehicle in specific situations. According to this broad approach, vehicle data not only includes 'vehicle internal data' (e.g. engine control unit data or electronic stability control unit data), but also 'vehicle external data' (e.g. the vehicles environment perceived by sensor units such as cameras or lidar sensors). Furthermore, 'user data', such as preferred audio settings or telephone contact lists can be considered as data being processed by the vehicles systems and control units.

2.2 Automated Driving

The definition of automated driving follows the SAE J3016 five respectively six level approach (Level 0 – Level 6) as shown in illustration 1:

- Level 0: No Automation / Driver Only
- Level 1: Driver Assistance
- Level 2: Partial Automation
- Level 3: Conditional Automation
- Level 4: High Automation
- Level 5: Full Automation / Autonomous Driving

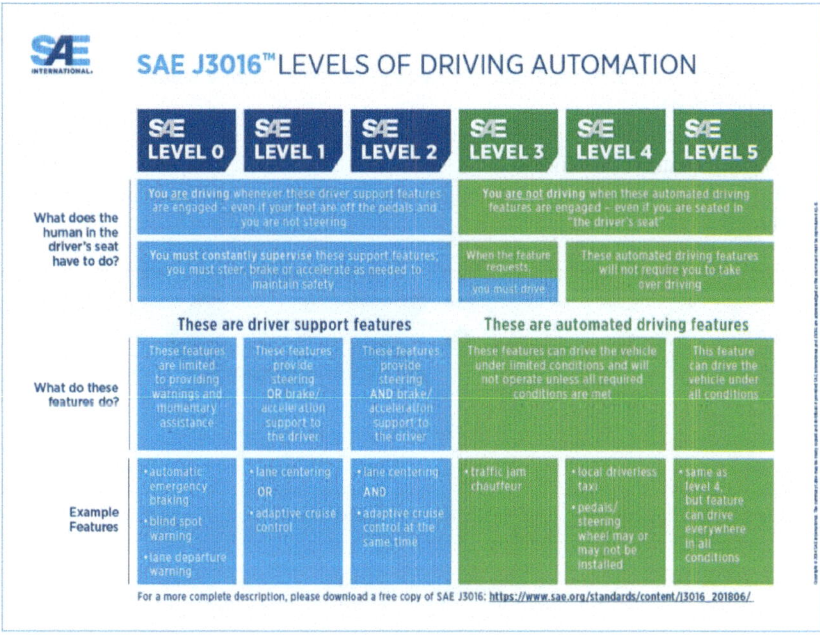

Illustration 1: Levels of driving automation (SAE J3016)[1]

Other approaches (e.g. BASt in Germany or NTHSA in USA) may vary in details but do not contradict this papers general statements.

2.3 Vehicle Lifecycle

The vehicle lifecycle can be considered as the complete timespan between first conceptual ideas and final dismantling of all produced vehicles of a specific vehicle model. After SOP (start of production) the size of the vehicle fleet continuously rises while it declines after EOP (end of production). For further understanding, following phases can be summarized as illustrated in illustration 2:

1 Source: https://www.sae.org/news/2019/01/sae-updates-j3016-automated-driving-graphic

- 'Research and Development' including prototypes, pre-production and type approval
- 'Manufacturing and Assembling' defined by SOP (start of production) and EOP (end of production)
- 'On Road' including all produced vehicles in the new-car market, used-car market, and export market activities.
- 'Dismantling and Recycling' including raw material extraction and spare parts distribution

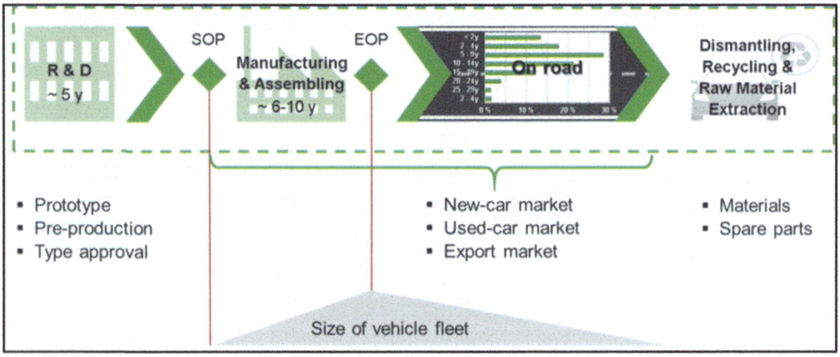

Illustration 2: Vehicle Lifecycle[2]

3 Connected Vehicles and Vehicle Data – Challenges and Opportunities

3.1 Connected Vehicles as Part of the Internet of Things

Modern vehicles from various OEM (e.g. Mercedes me, BMW connected drive, GM onstar) nowadays already are part of the 'internet of things' as they are connected to OEM-servers via embedded SIM card(s) and cellular networks as well as WiFi capabilities. In the end consumers focus and accelerated by media broadcast, this connectivity offers a wide range of benefits and opportunities as so-called 'over-the-air updates' (OTA) allow interactions between the OEM and their customers. For example, new enhanced features and functions may be offered to the customer in a post-sale phase (e.g. post-sale purchase of Tesla's autopilot function) as well as updating / upgrading navigation system map data (e.g. Mercedes me). Furthermore, connectivity

2 Source: KTI, Kraftfahrzeugtechnisches Institut und Karosseriewerkstätte GmbH & Co. KG

allows fixing software bugs and thus supports an optimization of time-to-market period during R&D. 'Firmware over-the-air' updates provide an additional option for further implementation of connectivity features.

On the other hand, connected vehicles are underlying the same potential risks as any other connected device (e.g. smartphones, smart TVs.) due to general cyber security issues. Table 1 summarizes potential risks and challenges in the categories 'vehicle safety', 'automotive security', and 'data privacy':

Table 1: Challenges and risks of connected vehicles

	Challenges	Risks
Vehicle Safety	Ensure active, passive and cyber safety	Physical injury or damage to health (in-/directly)
Automotive Security	Protection of vehicle functions and electronic components	Data manipulation and cyber attacks
Data Privacy	Driver / vehicle owner must be enabled to choose which data is shared	Driver / vehicle is not informed about data usage

3.2 Necessity for Data Access

Over the vehicles lifecycle different stakeholder groups gather for non-discriminatory access to 'vehicle internal data', 'vehicle external data', and 'vehicle user data' for many heterogeneous purposes. According to statements, level 3 vehicles generate up to 0,5 gigabytes (GB) of data per second on-board whilst the amount of generated data is to be expected to crank up to 1,5 GB/s for level 4 automation[3]. Due to limited processing power on the one hand side and limited cellular network capabilities (e.g. technical and financial restrictions) on the other hand side, only relevant sets of data can realistically be gathered and published for following exemplary interested parties:

- Vehicle Manufacturers (OEM)
- V2X
- Customer
- Other stakeholders

3 Source: ATZextra, Sonderheft August 2018, Big Data – Zwischen Fahrzeug und Backend, p.7

3.3 Stakeholder and Interested Parties

3.3.1 Vehicle Manufacturers (OEM)

The OEM's main interests can be considered to gain further and deeper information about the vehicles fleet condition with regard to (predictive) maintenance and service issues. Feedback on systems, functions, functionalities, and components regarding product improvement and a backloop towards R&D top off the technical perspective. With respect to a sales perspective, updates and upgrades of functionalities – that may also be limited in time regarding the usage of that enhanced functionalities – show a potential for significant revenue increase during the post-sale phase.

3.3.2 V2X

Vehicle-to-vehicle (V2V), vehicle-to-infrastructure (V2I), and vehicle-to-everything (V2X) connectivity not only imply the general possibility to exchange data between vehicles and other devices, it also describes the demand for specific sets of data. Possible use cases are intelligent traffic management systems, automated and autonomous driving tasks or weather alerts which all may require relevant data being generated by vehicles (e.g. the signal of a rain-light sensor for weather information) or perceived by the vehicle (e.g. traffic light status captured by a camera). Additionally, traffic accident reconstruction in case of an accident between vehicles with different degrees of automation may become a requirement.

3.3.3 Customer

Business customers and fleet operators as well as private vehicle users may ask for technical vehicle information being provided via an interface (e.g. an app or a web portal). In particular, uses cases vary from fuel consumption information, general vehicle and system status, a driver's log or geofencing functions (e.g. vehicle localization tool).

3.3.4 Other Stakeholders

Besides the already mentioned interested parties regarding vehicle data, a wide range of further stakeholders may have a demand for relevant and corresponding data. Insurance companies may be interested in the driving behaviour of the vehicle users in order to offer individualized insurance premiums as well as adapting vehicle generated or captured data for traffic accident reconstruction. Claims management organizations strive for information about damaged parts and components due to a vehicle break down or an accident. Road side rescue services may consider the vehicles geo-location as useful for enhancing their business services. Technical services and legal authorities need to consider the technical status of a vehicle, for example in case of a periodic technical inspection (PTI). Gas stations could enhance their services proactively, if they were informed about a vehicles fuel level in combination with the remaining range and route to be driven.

3.4 Interim Conclusion

Connected vehicles will contribute to the internet of things and will be capable to generate and perceive internal and external vehicle data as well as user data. This fact given allows market participants to enhance their individual business models respectively back up legal entities for their purposes. The foregoing statements shall point out the general opportunities of connected vehicle's data for individuals, market participants and society. Alongside, existing restrictions and risks have to be considered and request further analysis.

4 Vehicle Data Access Scenarios

4.1 Situation at Present

As previously analyzed and described, the vehicle embedded SIM card(s) directly communicate and exchange data between vehicles and corresponding OEM backend server platforms. According to individual usage agreements and in dependence of chosen value added services (between vehicle owner and vehicle manufacturer) a broad range of data is used to ensure services as contracted between the corresponding parties. This logical constellation between product provider (OEM) and customers shows significant benefits in terms of product liability and information technology security management with regard to cyber security. Apart from that, consequences in form of restrictions for economic market equilibriums already exist as access to generated data is strongly dependent on the vehicle manufacturers' willingness to share these data sets with other market participants (e.g. in the automotive aftermarket or the insurance industry).

4.2 Alternative Concepts

Besides the present situation (concept of an extended vehicle backend), alternative solutions are feasible offering a wide range of potential benefits for non-OEM service providers (e.g. insurers, independent repair shops) as well as public bodies and entities (e.g. courts, technical services, prosecutor attorneys, and police). Illustration 3 shows two alternative concepts including so-called 'third-party-servers', whether independently operated or not.

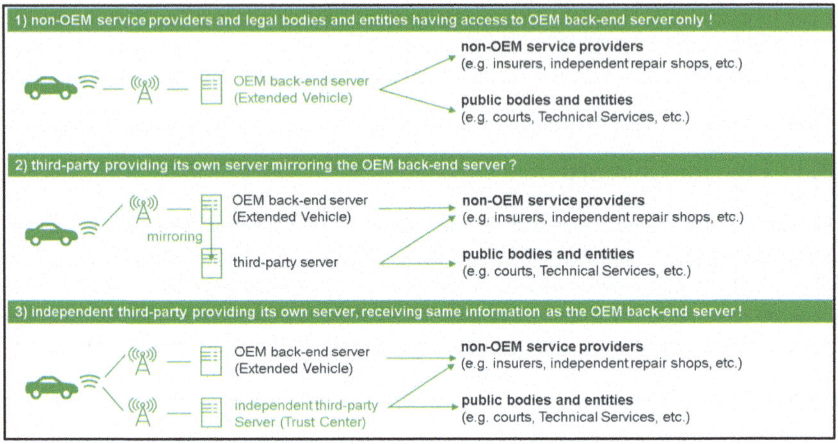

Illustration 3: Overview of Connectivity Concepts[4]

As the present conceptual situation, all alternative concepts also show disadvantages with respect to a wide range of criteria. Thus, further analysis and research has to be considered to elaborate a solution allowing all interested parties to perform their individual business models. Further solutions, for example 'black box' solutions and event data recorders are not within the focus of this article.

5 Outlook and Conclusion

This essay puts emphasis on the present and future challenges automated vehicles are being faced with. As complexity of the automotive industry in general and connected vehicles in particular arises, future concepts have to consider the holistic framework and market conditions from all different kinds of perspectives (e.g. the end consumer). Furthermore, not only a short-term consideration of vehicle usage is mandatory, instead the whole vehicles' lifecycle has to be taken into account including the long-range timespan after vehicles have been introduced to the market. The illustration of alternative connectivity concepts with regard to relevant vehicle data should sensitize all market participants to reassess existing concepts under a holistic perspective.

4 Source: KTI, Kraftfahrzeugtechnisches Institut und Karosseriewerkstätte GmbH & Co. KG

Haftungsfragen im Zusammenhang mit hoch- und vollautomatisierten Fahrzeugen

Dr. Philipp Ehring

T. Bertram (Hrsg.), *Automatisiertes Fahren 2019*, Proceedings,
https://doi.org/10.1007/978-3-658-27990-5_8

Autonom fahrende Autos erobern mehr und mehr unseren Alltag. Bereits jetzt sind viele Fahrzeuge mit Assistenzsystemen des Level 1 ausgestattet. Diese unterstützen den Fahrer, indem sie den Abstand zum vorausfahrenden Fahrzeug regeln, die Geschwindigkeit halten oder beim Spurwechsel vor einem Fahrzeug im toten Winkel warnen. In naher Zukunft soll das Auto aber mehr können. Durch Fahrassistenzsysteme soll nicht nur die Sicherheit des Straßenverkehrs erhöht werden.[1] Fast jeder dritte Autofahrer in Deutschland (31 Prozent) verbringt an einem durchschnittlichen Werktag mehr als eine Stunde in seinem Wagen; unter den Vielfahrern sind es sogar zwei Drittel (67 Prozent).[2] Diese Zeit lässt sich mit einem autonom fahrenden Wagen völlig anders nutzen. So können Emails bereits im Auto auf dem Weg zur Arbeit beantwortet, wichtige Papiere gelesen oder Kalkulationen überarbeitet werden. Das Büro wird mehr und mehr zur Kommunikationszentrale. Einzelbüros werden durch Großraumbüros ersetzt, Schreibtische weichen den Lounge-Areas, in denen nur noch besprochen und genetzt werkt wird. Damit diese Visionen Realität werden, bedarf es eines klaren Rechtsrahmens, der Verantwortlichkeiten definiert: Wer haftet, wenn das autonom fahrende Fahrzeug versagt, in einen Unfall verwickelt ist oder einen solchen gar verursacht?

I. Halter- und Fahrerhaftung

Erste Quelle für entsprechende Haftungstatbestände ist das Straßenverkehrsgesetz (StVG) mit der Halterhaftung nach § 7 StVG und der Fahrzeugführerhaftung nach § 18 StVG. Dieses hat der Gesetzgeber bereits Mitte 2017 an die Anforderungen des hoch- oder vollautomatisierten Fahrens angepasst und dieses nicht nur grundsätzlich für zulässig erklärt sondern auch besondere Pflichten für den Fahrer eines hoch- oder vollautomatisiert fahrenden Autos festgelegt.

Um sich den mit dieser Gesetzesänderung verbundenen Konsequenzen zu nähern, sind zunächst die verschiedenen Stufen des automatisierten Fahrens zu unterscheiden:[3]

Level 1: Fahren mit Assistenzsystemen
Level 2: teilautomatisiertes Fahren
Level 3: hochautomatisiertes Fahren
Level 4: vollautomatisiertes Fahren
Level 5: vollautonomes Fahren

Ganz am Anfang steht Level 0, bei dem das Fahrzeug nur durch den Menschen ohne Assistenzsysteme bewegt wird. Fahrassistenzsysteme unterstützen den Fahrer erst auf der

[1] *König*, in: NZV 2017, 123, 124.

[2] Bevölkerungsrepräsentative Umfrage „Auto-Emotionen" des Meinungsforschungsinstituts forsa im Auftrag von CosmosDirekt. Im Mai 2016 wurden in Deutschland 2.004 Autofahrer ab 18 Jahren befragt, die ein Auto im Haushalt besitzen. Als Vielfahrer wurden Fahrer mit einer jährlichen Fahrleistung von 20.000Kilometer oder mehr definiert.

[3] *v. Kaler/Wieser*, in: NVwZ 2018, 369.

Ebene des Level 1. Im Level 2 gehen diese Fahrassistenzsysteme so weit, dass ein teilautonomes Fahren möglich ist. Auf diesem Level wirken mehrere Assistenzsysteme zusammen, so dass in bestimmtem Umfang das Fahrzeug autonom fährt. Ein Beispiel hierfür ist der Stauassistent, bei dem der Abstandsregeltempomat und der Spurhalteassistent so zusammenarbeiten, dass das System in niedrigen Geschwindigkeitsbereichen selbständig beschleunigen, bremsen und lenken kann.

Level 3 beinhaltet das hochautomatisierte, Level 4 das vollautomatisierte Fahren. Nach der Legaldefinition des § 1 a Abs. 2 StVG müssen Fahrzeuge dieser beiden Stufen folgende Voraussetzungen erfüllen, um als hoch- oder vollautomatisiert zu gelten: Das Fahrzeug muss über eine technische Ausrüstung verfügen,

- die zur Bewältigung der Fahraufgabe – einschließlich Längs- und Querführung – das jeweilige Kraftfahrzeug nach Aktivierung steuern kann (Fahrzeugsteuerung),
- die in der Lage ist, während der hoch- oder vollautomatisierten Fahrzeugsteuerung den an die Fahrzeugführung gerichteten Verkehrsvorschriften zu entsprechen,
- die jederzeit durch den Fahrzeugführer manuell übersteuerbar oder deaktivierbar ist,
- die die Notwendigkeit (Erforderlichkeit) der eigenhändigen Fahrzeugsteuerung durch den Fahrzeugführer erkennen kann,
- die dem Fahrzeugführer das Erfordernis der eigenhändigen Fahrzeugsteuerung mit ausreichender Zeitreserve vor der Abgabe der Fahrzeugsteuerung an den Fahrzeugführer optisch, akustisch, taktil oder sonst wahrnehmbar anzeigen kann und
- die auf eine der Systembeschreibung zuwiderlaufende Verwendung hinweisen kann.

Vollautomatisierte Fahrzeuge unterscheiden sich von hochautomatisierten dadurch, dass innerhalb der anerkannten Systemgrenzen das Fahrzeug grundsätzlich selbst in der Lage sein soll, alle denkbaren Ereignisse im Straßenverkehr aus eigener Kraft zu bewältigen.[4]

Die höchste Stufe ist Level 5, das vollautomatisierte Fahren, bei dem kein Fahrer mehr notwendig ist.

Sowohl auf der Ebene des Level 1 als auch des Level 2 ist der Fahrzeugführer voll verantwortlich für das fahrende Auto und darf den Blick nicht vom Verkehrsgeschehen abwenden.[5] Er haftet damit neben dem Halter, der nach § 7 StVG (Halterhaftung) verantwortlich ist, nach wie vor aus § 18 StVG (Fahrzeugführerhaftung). Fahrzeugsysteme des Level 5 werden bislang nicht von der StVG erfasst, auch nicht von der Neuregelung des § 1 a StVG, so dass sie im Straßenverkehr nicht zulässig sind.

Halter von Fahrzeugen auf Level 3 und 4 unterliegen ebenfalls der verschuldensunabhängigen Gefährdungshaftung des § 7 StVG. Ist also ein hoch- oder vollautomatisiertes Fahrzeug in einen Unfall verwickelt, haftet sein Halter, gleich ob er an dem Unfall schuld ist oder das hoch- bzw. vollautomatisierte System den Unfall verursacht hat. Eine Anpassung der Halterhaftung hat der Gesetzgeber bislang abgelehnt.[6] Der Halter könnte sich allenfalls auf einen Fall der höheren Gewalt berufen (vgl. § 7 Abs. 2 StVG). Angesichts des gesetzgeberischen Willens, bei Unfällen mit automatisierten Fahrzeugen, die auf ein Systemversagen zurückgehen, die Frage der Haftung im Sinne des Unfallopfers zu klären, dürfte dieser Einwand aber erheblichen Bedenken unterliegen.[7] Entsprechend ist nicht zu erwarten, dass Gerichte künftig bei einem Systemversagen einen Fall höherer Gewalt akzeptieren.

[4] *Nugel*, in: AnwZert VerkR 10/2018 Anm. 2, S. 2.

[5] *Nugel*, in: AnwZert VerkR 10/2018 Anm. 2, S. 1.

[6] *König*, in: NZV 2017, 123, 126.

[7] BR-Drs. 69/17, S. 8.

Während die Halterhaftung bisherigen Grundsätzen entspricht, hat der Gesetzgeber mit der Neuregelung des StVG den Fahrern von hoch- und vollautomatisierten Systemen besondere Pflichten auferlegt (vgl. § 1 b StVG). Hierdurch hat er deren Verantwortlichkeit an die neuen technischen Möglichkeiten angepasst. Rechtstechnisch bedurfte es hierzu eines Kunstgriffs, da bei Zuschaltung der Systeme nicht mehr der Fahre das Fahrzeug lenkt, sondern das hoch- oder vollautomatische System. Entsprechend hat der Gesetzgeber in § 1a Abs. 4 StVG bestimmt, dass Fahrzeugführer auch derjenige ist, der eine hoch- oder vollautomatisierte Fahrfunktion aktiviert und zur Fahrzeugsteuerung verwendet, selbst wenn er das Fahrzeug nicht eigenhändig steuert.

Der Gesetzgeber geht aber noch einen Schritt weiter, und erlaubt diesen Fahrzeugführern ausdrücklich, sich während der Fahrt mit einem hoch- oder vollautomatisierten Fahrzeug vom Verkehrsgeschehen und der Fahrzeugsteuerung abzuwenden (§ 1 b Abs. 1 StVG). Allerdings müssen Fahrer solcher Fahrzeugsysteme wahrnehmungsbereit bleiben. Sie müssen reagieren können und die Steuerung des Autos unverzüglich, also sofort, wieder übernehmen, wenn das hoch- oder vollautomatisierte System hierzu auffordert. Reagieren muss der Fahrer aber auch dann, wenn er erkennt oder aufgrund offensichtlicher Umstände erkennen muss, dass die Voraussetzungen für eine bestimmungsgemäße Verwendung der hoch- oder vollautomatisierten Fahrfunktionen nicht mehr vorliegen (vgl. § 1 b Abs. 2 StVG). Der Fahrzeugführer muss daher insbesondere die in der Systembeschreibung aufgezeigten Grenzen für den Einsatz des hoch- oder vollautomatisierten Fahrsystems kennen und beachten. Beispielsweise muss er ein System, das nur für die Autobahn konzipiert ist, deaktivieren, wenn er die Autobahn verlässt. Der Fahrer eines hoch- oder vollautomatisierten Fahrzeugs muss also stets in der Lage sein, bei Vorliegen entsprechender offensichtlicher Umstände, die Entscheidung zu treffen, ob er die Fahrzeugsteuerung selbst übernimmt, auch wenn ihn das Fahrzeug hierzu nicht auffordert. Des Weiteren muss er in der Lage sein bei technischen Störungen (z. B. fahrzeugseitigen Warnungen) und sonstigen Störungen im Betrieb des Fahrzeugs (z. B. Reifenplatzer), die Steuerung wieder zu übernehmen.[8]

Die Neuregelung der StVG stellt damit an den Fahrzeugführer besondere Anforderungen. Da er gemäß § 18 Abs. 1 S. 2 StVG nur im Falle eines Verschuldens haftet, wird die Übernahmebereitschaft zukünftig in den Fokus bei Unfällen mit hoch- oder vollautomatisierten Fahrzeugen rücken. Da aber schon heute technisches Versagen regelmäßig zu einer Entlastung nach § 18 Abs. 1 S. 2 StVG führt, bleibt abzuwarten, ob dem Fahrer zukünftig ein Fahrlässigkeitsvorwurf gemacht wird, wenn ein Unfall durch einen Fehler der Fahrzeugautomatik verursacht wird und er im Rahmen seiner Übernahmepflicht nicht erkannt hat, dass das System überfordert war.[9]

II. Produkthaftung

Die Haftung des Herstellers eines hoch- oder vollautomatisierten Fahrzeugsystems hat der Gesetzgeber nicht ausdrücklich geregelt. Hierzu ist nach wie vor auf die allgemeinen produkt- oder produzentenhaftungsrechtlichen Vorschriften abzustellen. Bei diesen ist Anknüpfungspunkt für die Haftung das Inverkehrbringen eines fehlerhaften Produkts. Fehlerhaft ist ein Produkt, wenn es nicht den berechtigten Sicherheitserwartungen des Personenkreises entspricht, an das sich das Produkt richtet. Sowohl im Rahmen der Produkt- als auch der Produzentenhaftung treffen den Hersteller des Produktes Konstruktions-, Fabrikations-, Instruktions- und Produktbeobachtungspflichten. Zur Ausformulierung dieser Pflichten, wird man auf die vom Gesetzgeber vorgegebene Definition eines Fahrsystems als

[8] BR-Drs. 69/17, S. 16.

[9] *König*, in: NZV 2017, 123, 128.

hoch- oder vollautomatisiert (§ 1 a Abs. 2 StVG) zurückgreifen können.10 Wird nämlich ein Auto als hoch- oder vollautomatisiert verkauft, spricht einiges dafür, dass die Fahrzeugnutzer davon ausgehen dürfen, der Hersteller habe seine Konstruktionspflicht erfüllt und das Fahrzeug selber genügt den Anforderungen des § 1 a Abs. 2 StVG (vgl. oben).[11] Im Rahmen der Instruktionspflicht wiederum wirkt sich der Umstand aus, dass ein hoch- oder vollautomatisiertes Fahrzeug nur im Rahmen des bestimmungsgemäßen Gebrauchs genutzt werden darf. Der Hersteller muss daher den von ihm vorgesehenen Einsatz und den Grad der Automatisierung gegenüber dem Fahrzeugnutzer kommunizieren. Darf beispielsweise ein Kraftfahrzeug nicht mittels einer automatisierten Fahrfunktion auf Landstraßen betrieben werden, weil das System nur für den Einsatz auf Autobahnen vorgesehen ist, muss die Systembeschreibung des Fahrzeugs hierüber unmissverständlich Auskunft geben.[12] Tut sie das nicht, verletzt der Hersteller seine Instruktionspflicht und macht sich haftbar.

III. Black Box

Nach alldem wird zukünftig ein Gerät in hoch- oder vollautomatisierten Fahrzeugen Bedeutung erlangen, das bislang hauptsächlich aus Flugzeugen bekannt ist, die Black Box. Der Gesetzgeber hat hierzu § 63 a StVG neu in das Gesetz eingefügt, der die Notwendigkeit und den Umgang mit den in der Black Box gesammelten Daten regelt. Aufgabe der Black Box ist es, im Falle eines Unfalls genau nachzuvollziehen, wer gefahren ist, der Computer oder der Fahrer und ob der Fahrer zum Eingreifen aufgefordert wurde. Die hierzu gesammelten Daten werden sechs Monate gespeichert, anschließend sind sie zu löschen.

IV. Selbstlernende Systeme/IT-Sicherheit

In Grenzbereiche stoßen heutige Rechtssystem hingegen vor, wenn das automatisierte Fahrzeug selber lernt (sog. Machine Learning). Produkthaftung geht davon aus, dass die Inverkehrgabe eines Produkts eine dem Hersteller zurechenbare Gefahreröffnung darstellt. Haftbar ist also immer der Hersteller als natürliche oder juristische Person und nie, wie es bisweilen diskutiert wird, das autonome System. Dies führt dazu, dass der Hersteller auch weiterhin haften muss, obwohl sich das selbstlernende System aufgrund seiner Lernerfahrung verändert und womöglich nicht mehr dem Produkt entspricht, das der Hersteller ursprünglich in Verkehr gebracht hat. Gesteigert wird diese Problematik, wenn unberechtigte Eingriffe von Außen (sog. Hackerangriffe) auf das System zugreifen und es verändern. In all diesen Fällen wird man auch zukünftig die oben beschriebenen produkthaftungsrechtlichen Mechanismen anwenden und die Konstruktions- bzw. Produktbeobachtungspflicht des Herstellers zum Anknüpfungspunkt seiner Haftung machen.13 Der Hersteller hat damit die selbst lernenden Komponenten des automatisierten Fahrzeugs so zu konstruieren, dass die berechtigten Sicherheitserwartungen der Fahrzeugnutzer nach dem Maßstab des § 1 a Abs. 2 StVG auch weiterhin erfüllt werden und auch nicht durch sog. Hackerangriffe beeinträchtigt werden können. Weiterhin muss der Hersteller im Rahmen seiner Produktbeobachtungspflicht die Lernerfahrungen des automatisierten Fahrzeugs kontrollieren und so sicherstellen, dass das Gelernte vom System nicht in einer Weise gespeichert wird, die nicht mehr den Sicherheitserwartungen der Fahrzeugnutzer entspricht.

[10] *Armbrüster*, in: ZRP 2017, 83, 84; *von Bodungen/Hoffmann*, in NZV 2018, 97, 101.

[11] *von Bodungen/Hoffmann*, in: NZV 2018, 97, 99.

[12] BR-Drs. 69/17, S. 13 f.

[13] vgl. *Hartmann*, in: PHi 2017, 2, 8.

V. Fazit

Insgesamt hat der Gesetzgeber mit wenigen neu in das StVG eingefügten Paragraphen die Haftung für Unfälle durch hoch- oder vollautomatisierte Fahrzeuge umfangreich geregelt. Damit hat er die Rechtssicherheit geschaffen, die Ingenieure benötigen, um das umzusetzen, was technisch möglich ist. Vorgaben zum vollautonomen Fahren fehlen allerdings bisher: hier bleibt abzuwarten, was die technische Entwicklung in den nächsten Jahren hergibt und wie dies durch den Gesetzgeber begleitet wird. Aus haftungsrechtlicher Sicht wird es zukünftig unerlässlich sein, dass Konstruktionsleistungen im Bereich automatisierter und autonomer Fahrsysteme frühzeitig – bereits im Entwicklungsstadium – juristisch überprüft werden. Es gilt nämlich nicht nur das, was technisch möglich ist, rechtssicher umzusetzen. Eine rechtssichere Konstruktion automatisierter und autonomer Systeme erfordert auch, dass das, was dem neusten Stand von Wissenschaft und Technik entspricht, zum Einsatz kommt. Nur so können haftungsrechtliche Risiken sicher vermieden werden.

Allocation of liability costs between motor insurers and vehicle manufacturers – an analysis of the current liability and insurance framework for automated vehicles

Fabian Pütz

T. Bertram (Hrsg.), *Automatisiertes Fahren 2019*, Proceedings,
https://doi.org/10.1007/978-3-658-27990-5_9

1 Introduction

The progressive technical development in the field of advanced driving assistance systems (ADAS) and automated driving makes it feasible that motor vehicles with a high level of automation will be ready for market within the next decade. The vision of automated vehicles on public roads not only depends on technical progress but also requires an adequate liability framework to ensure legal certainty for the relevant stakeholders on both the demand and supply side.

Several national legislators have recognized the need for adjustment and are currently discussing amendments to existing liability laws. The German legislative body has already implemented an amendment to the existing Road Traffic Act (Straßenverkehrsgesetz (StVG)) in June 2017. This amendment sets legal requirements for the operation of highly and fully automated vehicles[1] and mainly preserves the existing liability and insurance framework, characterised by a combination of strict liability of the vehicle owner[2] and fault-based liability of the (human) driver. This liability framework is coupled with obligatory insurance to be maintained by the owner covering liability costs of himself, the driver and the titleholder of the respective vehicle. OEMs are not directly included into the legal framework and the legislator expects that motor insurance companies examine and conduct subrogation claims under existing product liability law in the case that a defect of the automated driving system causes an accident.

In general, a functioning liability and insurance framework should meet two central objectives. In the external relationship to the damaged third party, the design of the framework has to ensure an adequate level of claimant protection. In the internal relationship between the liable parties, a reasonable and adequate ultimate allocation of liability costs fulfils an important incentive function for the involved parties. Because the progressive automation of motor vehicles will successively reduce the relevance of the owner and human driver for the occurrence of road accidents, we use the liability and insurance framework resulting from the amendment to the StVG in Germany as a case-study to analyse whether this framework adequately ensures these central objectives.

1 The German legislative body does not follow the classification wording of SAE but uses the terminology of the German Federal Highway Research Institute. The equivalent levels of automation referring to SAE nomenclature would be conditional automation (level 3) and high automation (level 4).

2 We use the term "owner" for the German legal term "Fahrzeughalter" and the term titleholder for the German legal term "Fahrzeugbesitzer". German law distinguishes between the right of property title and the right of disposal. The owner of a vehicle is the person, who owns the right of disposal and has the duty to maintain the vehicle in safe condition. The titleholder is the person or institution formally owning the property title of the vehicle.

2 Current law regarding legal liability for automated vehicles

Due to the progressing automated driving capability of motor vehicles, the German legislative body has recognized the need for adjustments to the StVG. Therefore, an amendment was elaborated which came into force in June 2017[3]. The aim of this amendment is to ensure a clear allocation of liability for highly and fully automated vehicles (SAE level 3 and 4). Autonomous[4] vehicles are not within the scope of this amendment yet.

In general, the StVG distinguishes between two parties, which can be addressed when a third party suffers damage due to the operation of a motor vehicle. While the StVG imposes strict liability on the owner of a vehicle (§ 7 StVG), the human driver is only liable based on fault (§ 18 StVG). This legal approach is coupled with the obligation for the owner to take out and maintain motor third party liability (MTPL) insurance coverage protecting himself, the titleholder and the driver of the vehicle against liability claims from third parties caused by the use of the vehicle (§ 1 Law on compulsory insurance for owner s of motor vehicles (PflVG)). In this respect, the damaged third party has the possibility to directly claim against the respective MTPL insurer (§ 115 Insurance contract law (VVG)).

2.1 Liability of the owner of the motor vehicle

The strict liability approach of the German liability system is based on the legal logic that the usage of a motor vehicle opens an (abstract) source of risk to society. This exposition of risk is legally accepted but the individual, who benefits from the usage of the risky object, has to bear strict liability if the (abstract) risk becomes manifest. Next to owners of motor vehicles the concept of strict liability is *inter alia* applied to owners of aircraft, animals and operators of railways and power plants.

The owner does not necessarily need to be the titleholder of the vehicle, as German law separates the right of property title determining the titleholder (§ 903 German Civil

3 Achtes Gesetz zur Änderung des Straßenverkehrsgesetzes vom 16. Juni 2017 (Bundesgesetzblatt Jahrgang 2017 Teil I Nr. 38 Seite 1648).

4 We use the term "autonomous" vehicle for vehicles which technically have the capability to self-reliantly fulfil the driving task in a comprehensive manner and do not require a (human) driver at all (SAE level 5).

Code (BGB)) and the right of possession determining the owner (§ 854 BGB)[5]. The main characteristics of the owner are that he is the person, who uses the vehicle for his own account and has the power of disposition. The owner regularly bears the operating costs and benefits from the use of the respective motor vehicle[6].

The owner is obliged to compensate for claims with strict liability (§ 7 (1) StVG). To ensure the ability to meet these financial obligations, the owner has to maintain obligatory MTPL insurance coverage (§ 1 PflVG). The strict liability of the owner for losses due to the same event is generally limited to € 5 million for personal injury and death of a third party and to € 1 million for property losses (§ 12 (1) StVG). However, these maximum liability limits are doubled, if the damages result from the use of a highly or fully automated driving function (§ 12 (1) StVG).

2.2 Liability of the driver of the motor vehicle

According to the StVG, the driver of a vehicle is liable up to equal compensation limits as the owner, if damage to a third party is due to his fault. For this, the fault of the driver is presumed (§ 18 StVG) but can be refuted by providing evidence.

In the amendment of the StVG, the definition of the "vehicle driver" is not extended to the automated system but designates the human driver as the driver of the vehicle even for the time he does technically not control the vehicle by an active steering action (§ 1a (4) StVG). Within the scope of the intended use of the automated driving function, the driver of the vehicle is generally allowed to turn away from the traffic scene and vehicle control (§ 1b (1) StVG), but has to keep situational awareness to always fulfil his obligation to immediately (without undue delay) take over vehicle control

- if the highly or fully automated system so requests or (§ 1b (2) no. 1 StVG)
- if he recognizes or should recognize on the basis of obvious circumstances that the conditions for the intended use of the highly or fully automated driving functions no longer exist (§ 1b (2) no 2 StVG).

Even if the amendment to the StVG does not explicitly differentiate between highly and fully automated vehicles, the legal requirements of the driver of the vehicle implicitly lead to a certain distinction. This is because not only the scope of the intended use but

5 For example, in case of a credit-financed car the bank typically is the titleholder of the car as long as the debtor has not paid back the credit. However, the debtor still has the factual right of daily disposition and the duty to adequately maintain the vehicle as the owner of the vehicle.

6 Dorothee Euler and Andrea Kreuter-Lange, „Kraftfahrzeughaftpflichtversicherung", in Halm, Engelbrecht and Krahe (eds.) Versicherungsrecht-Handbuch des Fachanwalts 4th edition (2011), at p. 1716.

also the duration and extent of (legally accepted) *turning away* from the driving scene should increase with a progressing automation.

In addition, the driver is also liable under the general provisions of the German Civil Code, if he intentionally or negligently, unlawfully injures the life, body, health, freedom, property or another right of another person (§ 823 (1) BGB). The conditions of an intentional injury are that the damaging party acts consciously and willingly with regards to the occurrence of the damaging event. A person´s act is classified as being negligent if the person fails to exercise reasonable care (§ 276 (2) BGB).

In contrast to the provisions of the StVG, the liability resulting from the BGB is not limited to a maximum amount. However, the damaged third party bears the burden of proof that the preconditions of liability (intentional or negligent, unlawful action) and the causality between the action and the damage are met.

2.3 Product liability

The vehicle producer is not explicitly included in the liability framework of the StVG. However, even if the car manufacturer is not explicitly mentioned as a liable person, the explanatory memorandum of the amendment to the Road Traffic Act clarifies that the respective motor insurance company has the possibility to subrogate claims under applicable product liability law.

As the provisions regarding product liability resulting from *Directive 85/374/EEC* are largely harmonized in the European Union, the following analysis focusses on the specific German implementation of these provisions in the Product Liability Act (ProdHG), acknowledging that producer liability can also result from either a contractual relationship or tortious claims under the German Civil Code.

The provisions of product liability hold a producer liable for claims of a damaged third party if a defective product causes a person's death, bodily injury, damage to his health or damage to an item of property (§ 1 (1) ProdHG). The liability resulting from ProdHG is strict.

The scope of the ProdHG only covers property damage if this damage was caused to an item of property other than the defective product (ibid). Following from this, the scope of the ProdHG does not cover spreading defects which are defects to a functionally separable component part that lead to a damage to the whole product. Hence, subrogation claims for damages to property items are only possible for MTPL insurance losses and not for insured losses to the own vehicle.

Furthermore, the producer has only to compensate for damages, if the item of property is of a type which is ordinarily intended for private use or consumption and mainly used accordingly by the damaged person (ibid). This exclusion is especially relevant for third party damages to commercially or publicly used property/infrastructure, such as commercial vehicles (buses, commercial trucks), buildings or traffic infrastructure, which are not covered by the literal interpretation of this provision. Assessing the wording of this legislation in the case of a car-to-car-accident event, vehicles mainly used for a commercial purpose would generally not be covered by the scope of the ProdHG. Today, prevailing possession-based mobility highly reduces the relevance of this exclusion. Assuming that progressive automation will promote a shift towards the usage of commercial shared-mobility services, the relevance of this exclusion could potentially grow in the future.

The ProdHG defines a product as defective when it does not provide the safety that the consumer, with consideration of all circumstances, is entitled to expect. Relevant circumstances which have to be considered are the presentation of the product, the kind of use that is reasonably expectable and the time the product was put into circulation (§ 3 ProdHG). For the legal interpretation of the existence of a defect, the objective safety expectations of a reasonable average user or consumer are used as a reference. The damaged party bears the burden to prove the defect, the damage and the causal relationship between the defect and damage (§ 1 (4) ProdHG). As an easing of the burden of proof, the damaged party only has to prove that the defect existed at the time of the loss occurrence, not the temporal and local origin of the defect[7]. If the damaged party can provide this evidence, the manufacturer has to prove that exclusions of liability are applicable.

In contrast to the provisions of the StVG, the ProdHG does generally not limit the maximum liability amount for damages caused to an item of property. However, the maximum liability amount for personal injury damages caused by identical products with the same defect is limited to € 85 MM (§ 10 ProdHG). The application of this maximum liability amount is not harmonized within the European Union but if a member state decides to implement a different liability limit, the minimum threshold of € 70 MM has to be applied[8]. Even if we assume that this limitation is high enough to cover losses arising from single accident events, this limitation of liability could especially gain in relevance, because the interconnection of vehicles and the correlation of risk behavior of a vehicle fleet based on the same algorithm potentially induces accumulation risk, where multiple accidents could cause personal injury damages in excess of the implemented threshold. In this case, it would be unreasonable if exceeding loss amounts would shift back to either the insured owner or the damaged third party.

7 A. Komescher, "Produkthaftung", in Halm, Engelbrecht and Krahe (eds.) Versicherungsrecht-Handbuch des Fachanwalts 4th edition (2011), at pp. 2229.

8 Council Directive of 25 July 1985 on the approximation of the laws, regulation and administrative provisions of the Member States concerning liability for defective products (85/374/EEC), art. 16.

Furthermore, the obligation to compensate for property damage under the ProdHG only applies if the respective damage exceeds a threshold of € 500 (§ 11 ProdHG). Hence, motor insurers could not subrogate property losses that do not exceed this threshold. Under current regulations, this is understandable, as legal and procedural costs induced by a conduction of subrogation claims could easily exceed this threshold and lead to economic inefficiencies. We expect that this threshold should not lead to a material inadequate allocation of liability costs from a macro-level view on the overall insured portfolio, as MTPL losses below this threshold account for only 0.43 % (2015)[9] of all insured MTPL loss expenditure. However, from the perspective of the individual insured person, not being able to conduct a subrogation claim because of the minimum threshold can have adverse consequences because of the impact to the individual insurance premium, which is commonly based on a no-claims bonus system in the German market.

In addition to the limitation of maximum liability amounts for personal damages and the deductible for property damages, the liability of the manufacturer *inter alia* is generally excluded, if under consideration of all circumstances it is probable that the defect which caused the damage did not exist at the time when the producer put the product into circulation (Article 1 (2) no. 2 ProdHG). This exclusion is relevant as the capability of automated vehicles will likely not be static but could be altered by (over-the-air) software updates so that the inherent accident risk exposure fluctuates over time. However, the provisions of the ProdHG generally do not oblige the producer to monitor the circulated product after the sale. Therefore, it should be discussed whether the temporal attachment point of putting the product into circulation has to be renewed with every (material) change to the product functionality. On a technical side, this discussion has also to be supported by vehicle approval authorities with setting guidelines defining material changes to system functionality, which require a repeat process of type approval test. If the temporal attachment point of product circulation would not be renewed with every (material) change, residual technological development risk would fall back to the owner of the vehicle, who is liable with strict liability.

In addition, the legal provision that claims based on the ProdHG expire ten years from the time when the producer put into circulation the product which caused the damage (Article 13 ProdHG) could also lead to a shift of residual technological risk to the owner of the vehicle. This exclusion will not be relevant in the first years with automated vehicles on the road but in case that an automated vehicle, which is older than ten years, causes an accident.

9 Share of loss expenditures for MTPL-losses, which do not exceed the threshold of € 500, from total loss expenditures for MTPL-insurance in the year 2015. Source: German Insurance Association.

Finally, the liability of the manufacturer is excluded if the state of scientific and technical knowledge at the time when the producer put the product into circulation was not such as to enable the detection of the defect (Article 1 (2) no. 5 ProdHG). To apply this exclusion of liability it has to be objectively referenced against the scientific and technical knowledge which was generally accepted and available at this time. This exclusion is potentially important, with the progressing automation of modern vehicles because general standards and sufficient case law decisions for programming driving algorithms and testing autonomous vehicles do not exist yet.

3 Adequacy of the existing liability and insurance framework for autonomous vehicles

Based on the description of the different sources of liability of the owner and driver of a motor vehicle on the one and the producer on the other side, the resulting liability and insurance framework is depicted in Figure 1.

Even if a claim of a damaged third party against the producer is legally possible, the dashed line reflects the associated higher financial and organisational effort required. Hence, the damaged third party will rather not use this path if the claim is sufficiently compensated by a direct claim against the motor insurer due to strict liability of the owner.

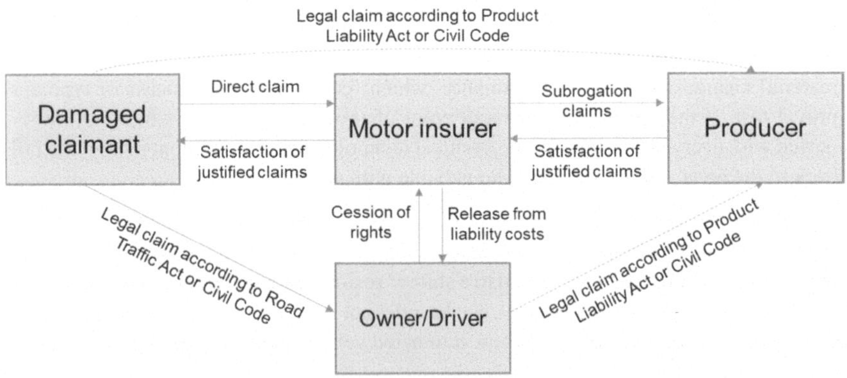

Figure 1: German liability and insurance framework for losses caused by (highly and fully automated) motor vehicles. The figure depicts an overview of the several sources of liability of the involved parties in case of losses due to road accidents.

Today, above 90 % of all road accidents are caused by human error. Therefore, an investigation of the product liability claims is very low in the (German) motor insurance market. However, with a progressive automation of motor vehicles and a shifting responsibility for the causation of accidents to OEMs and their suppliers, the interdependencies of the current liability and insurance framework and regulations regarding product liability have to be assessed and aligned to create a consistent and adequate framework for automated vehicles.

In the following, we analyse the adequacy of the existing liability and insurance framework for automated and autonomous vehicles based on the following criteria of assessment:

- **Protection of the claimant:** Does the current framework adequately ensure the protection of the damaged (third party) claimant?
- **Responsibility-based allocation of costs:** Does the current framework adequately allocate liability costs to the responsible party?
- **Cost-efficiency of the claims process:** Does the current framework ensure a cost-efficient allocation of liability costs?

3.1 Protection of the claimant

The central idea of the existing (strict) liability and insurance framework is to ensure a high as practicably possible level of protection for the damaged third party claimant. Claimant protection must be evaluated as a two-stage process considering the maximum extent of (financial) protection and the required processual effort to achieve this compensation. Hence, the following analysis assesses the easiness of the claims process (e.g. regarding the speed of the process, clarity of relevant institutions to contact and the burden of proof) and the material (maximum) compensation of the claimant.

The central idea of claimant protection from a processual point of view is anchored with the strict liability of the vehicle owner, obligatory MTPL insurance and the possibility to direct claim against the respective motor insurer. The strict liability of the owner and the possibility to claim directly against the motor insurance company is a generally proven process that ensures a clear path for compensation from the motor insurer as the central institution, which comprehensively compensates justified liability claims for all probably liable parties. In fact, this strict liability approach is also practicable with automated vehicles, so that it can serve as a blueprint for other countries′ legal approaches, which are based on fault of the human driver[10]. That said, the strict liability approach

10 Sophia Duffy and Jamie Hopkins, "Sit, Stay, Drive: The Future of Autonomous Car Liability", (2016) SMU Sci. & Tech. Law Rev. 101, at p. 116.

has to be embedded into a well-functioning and robust insurance market to effectively protect third party claimants.

The preservation of the current process for automated vehicles in large part will assure claimant-friendly processing of liability claims with respect to the speed of the process and clarity about the relevant institutions. This is supported by legislation, which generally limits the period for processing claims to three months (§ 3a PflVG). Therefore, a speedy processing of third party claims is supported by the legal environment. However, the adequacy of the current framework for autonomous vehicles will only guarantee a high level of claimant protection, if the maximum material compensation covered by this process is also adequately determined.

On the one hand, it could be argued that the current framework generally does not reduce the compensation of a claimant damaged by an automated vehicle. This is because the doubling of the maximum liability limits for owner´s strict liability increases the maximum achievable compensation that can be claimed without providing evidence of fault. The legislator *inter alia* rationalizes the need for an increase, because the fault-based unlimited liability of the (human) driver may not apply in the case that the accident is caused by a technical failure of the automated driving function. That said, it is questionable, whether the increase of maximum liability amounts will adequately replace the unlimited fault-based liability of the (human) driver and adequately ensure the protection of the claimant for severe loss scenarios. This is because the current approach of applying the liability of the owner and (human) driver, which are covered by motor insurance, on a first level and the possibility to subrogate claims under product liability law only on a second level procedurally separates the damaged third party from claims against the manufacturer. This separation could turn out to be disadvantageous if the claimant is only compensated up to the lower limitations of strict liability of the owner, even if the manufacturer is held liable during the subrogation claim process of the motor insurer.

With a limited financial tail-risk of single insured MTPL losses on a portfolio-basis, we expect that the great majority of accidents will be sufficiently covered by the increased maximum liability amounts. However, single accident occurrences have already shown that possible maximum losses for MTPL insurance can greatly exceed the applicable threshold. For instance, this can be illustrated by the highest recorded MTPL insurance loss, where due to the fault of the driver, a vehicle crashed with a train, which derailed and crashed into another train. Ten people died and about 80 were injured leading to an insured loss of about € 30 MM.

If this accident would have been caused by an automated vehicle, the claimant would (additionally) have to claim the manufacturer under applicable product liability laws for losses, which exceed the maximum compensation amounts of the StVG. This would expose the claimant to higher legal costs and effort. In turn, this potentially hinders claimants from claiming for compensation that exceed the maximum liability limits of the StVG, even if those claims would be legally justified. In doing so, the current framework could contravene the principle of a one-stop shop approach and hence reduce the level of claimant protection with regards to the required processual effort to claim for adequate material indemnification.

Therefore, policy sets may be improved by a further increase of maximum liability amounts for owner´s strict-liability. This approach is also imperative from a society´s point of view because the limitation of maximum liability amounts would otherwise potentially result in disproportionate residual legal risk and uncertainty of compensation to damaged third parties. From an insurance market point of view, this should not be problematic as commonly sums insured in the German market highly exceed the liability limits of the StVG[11].

3.2 Responsibility-based allocation of liability costs

With automated vehicles on the road, the responsibility for the causation of road accidents will successively shift from the human driver to the automated system itself. Therefore, only accidents that are caused by inadequate maintenance (e.g. missed software-updates) of the automated vehicle, can be attributed to the owner´s sphere of responsibility[12]. With autonomous vehicles, which do not need a human driver anymore, fault of the human driver will not be a relevant contributing factor to accident occurrences anymore. The owner of the vehicle will only be left with the possibility to influence the risk of the vehicle through a reasonable effort to maintain the physical components of the vehicle and update the autonomous system´s software. Hence, it is expected, that the share of accidents which potentially fall under the applicability of product liability will increase[13].

That said, the preservation of the strict liability approach for automated vehicles is systemically consistent with the legal logic that the owner of the vehicle, from whose use the owner benefits, opens a source of risk to society. Hence, the owner of the vehicle will ultimately bear the residual risk of liability, which legally cannot be attributed to the manufacturer or other potentially liable parties. However, the owner´s only limited possibility to influence the risk exposure of the vehicle makes it necessary to reconsider the balance of liability cost allocation between the owner and manufacturer of the vehicle in favor of the owner.

11 Commonly sums insured in the German motor insurance market are limited to € 100 million.

12 Andrea Bertolini et al., "On Robots and Insurance", *International Journal of Social Robotics* 381, at p. 385.

13 Bericht der Ethik-Kommission automatisiertes und vernetztes Fahren (Juni 2017), at p. 11.

From a macro-level view, which does not consider the allocation of liability costs after one specific insured loss event, it could be argued that the owner of the vehicle will ultimately carry liability costs anyway, either directly by paying insurance premiums or indirectly through an additional price component when purchasing a vehicle. This price component then accounts for the product liability risk of the manufacturer[14]. However, this argument only ensures that liability costs are adequately attributed to a superordinate entire portfolio (e.g. homogenous portfolio of insured or portfolio of buyers of certain vehicles) but not that liability costs are also adequately allocated on an individual basis (e.g. impact on no-claims discount on individual insurance contract).

Furthermore, this perspective does not consider the allocation of expense inflation risk to the owner or manufacturer. In case that the owner pays an additional premium when purchasing the vehicle, the risk that the premiums are not sufficient to meet liability claims is shifted to the OEM. By contrast, if the owner of the vehicle is still obliged to pay obligatory motor insurance premiums, the risk of expense inflation indirectly remains with the owner of the vehicle, as it iteratively flows back via insurance premium adjustments. In addition, allocating liability costs to the producer, who adds these costs as an additional premium to the vehicle´s price, would *ceteris paribus* induce that less reliable vehicles are in trend more expensive, which in turn reduces the marketability of unreliable vehicles. In this way, the allocation of liability costs to the producer would support a more prudent testing approach of manufacturers before placing their vehicles on the market[15].

It could further be argued, that a responsibility-based allocation of liability costs for a specific accident event is generally applied by the current framework because the motor insurer covering owner´s strict-liability in the external relationship to the damaged third party can subrogate claims from the producer under applicable product liability laws. In case that this subrogation claim is successful, the liability costs are ultimately born by the producer. In doing so, the design of the liability and insurance framework has to ensure that the residual risk of owner´s liability is not inadequately increased by the design and conditions of the subrogation claim processes. For this, the extent of reasonable and adequate allocation of liability costs depends on the willingness and the ability of the motor insurer to conduct subrogation claims under the current product liability laws.

The current scope of MTPL insurance is only limited to the satisfaction of justified claims and the rejection of unjustified claims ("passive legal protection", see § 2 (1) Regulation on insurance cover in motor vehicle liability insurance (KfzPflVV)). In contrast, MTPL insurance does not cover the owner´s interest in law enforcement against

14 See supra note 8.

15 Marchant and Lindor, "The Coming Collision Between Autonomous Vehicles and the Liability System", (2012) *Santa Clara Law Review* 1321, at p. 1337.

other parties ("active legal protection"). If a motor insurer settles a justified MTPL claim and the insured is entitled to reclaim damages from another third party, the claim of the insured is assigned to the motor insurer to the extent that the motor insurer already compensated the third party claimant ("cession of right", see § 86 VVG). Hence, the insured currently neither has a legal nor contractual right within the applicable law (§ 2 (1) KfzPflVV) and standard motor insurance wording that would oblige the insurer to conduct a subrogation claim against the vehicle producer or automotive supplier. This is not problematic for manually driven vehicles because subrogation claims to third parties do not play a significant role and do not affect the overall achievable premium volume of the motor insurance market. Thus, conducting subrogation claims against third parties is in the best (economic) interest of both the insured and motor insurer, as this prevents adverse impacts to the individual insurance contract of the insured and also lowers claims expenses and, in turn, improves underwriting results of the insurer. This alignment of interest could decrease because the increasing relevance of subrogation claims successively erodes the presence of insurance risk, which is a crucial precondition of the motor insurance business model. Given the market-wide importance of motor insurance business volume, this potential conflict of interest is especially relevant in case that a motor insurance company has not proactively initiated a structural change to reduce the dependency on contribution margins from MTPL insurance for the financing of companies´ overhead costs. Hence, assuming that a progressive shift of liability costs on a superordinate portfolio or market level decreases the overall achievable premium volume, the motor insurer may be less likely to conduct subrogation claims.

In addition to the willingness of the insurer to conduct subrogation claims, the ability of an insurer to do so depends on the legal and technical know-how and the availability of relevant (telematics) data. These resources are required to be able to provide evidence that the conditions of product liability are met. The burden of proofing the presence of a product defect could be difficult, especially for accidents caused by software failure of an automated vehicle. This problem is increased by the current limited technical and engineering know-how, which is required and therefore maintained by insurance companies to underwrite retail motor insurance business today. This could lead to the outcome that a subrogation claim against an OEM is not conducted due to a lack of technical expertise and the unknown probability of success. This, in turn, could lead to an inadequate allocation of liability costs at the expense of the insured owner of the vehicle.

Furthermore, the possibility of the motor insurer to conduct subrogation claims against the manufacturer of the vehicle is also limited, if the potential exclusions of product liability described in section 2.3 are applicable. This especially holds true for damages to the own vehicles, which are generally not covered by the provisions of the ProdHG.

3.3 Cost-efficiency of the claims process

Regardless of the question of the current framework´s ability to allocate liability costs for accidents in a responsibility-based manner, the owner of the vehicle and even producers are indirectly exposed to adverse economic consequences, if the process of allocating liability is not structured efficiently and unnecessarily causes additional processual and legal costs.

Generally, the combination of strict liability with mandatory MTPL insurance and the opportunity to direct claim against the motor insurer is appropriate to satisfy claims of a third party arising from manually driven vehicles with low processual costs. However, the efficiency of the current framework´s claims process is mainly based on the fact, that the owner of the vehicle and the driver are ultimately almost exclusively liable for third party losses caused by the usage of the insured vehicle. Therefore, the claimant directly contacts the respective motor insurance company as the institution, which also ultimately bears liability costs. Further litigation and subrogation claims[16] are only rarely conducted by the motor insurer after the direct claim process of the damaged third party[17].

However, the idea that the motor insurer is the central institution to claim and to absorb liability costs holds true for manually driven vehicles, where more than 90 % of all accidents are caused by human error, but this could change for higher automated vehicles if product liability becomes more relevant. Hence, even where liability costs can adequately be allocated through subrogation claims of the motor insurance company after a single accident event, the owner of a vehicle maintaining MTPL insurance still has to pay for internal overhead-costs (e.g. claims-handling, execution of the subrogation claim) and safety-margins for residual legal risk. These are accounted for in the motor insurance premium.

From a macro-level perspective of both the motor insurer and producer, the two-step approach of paying the claim of the damaged third party first and finally allocating liability costs by the way of subrogation claims potentially causes significant additional legal costs, which have to be absorbed by either the producer or motor insurance company. Ultimately, these costs are carried by the consumer either as the insured or purchaser of a car. Hence, an unaltered preservation of the current framework could turn out as inefficient, at the latest with autonomous vehicles on the road. Therefore,

16 Today, subrogation claims in the motor insurance business are mainly relevant for cases where one of the insured parties breaches contractual obligations that have to be fulfilled before or after a loss occurrence.

17 Carrie Schroll, "Splitting the Bill: Creating a National Car Insurance Fund to Pay for Accidents in Autonomous Vehicles" (2015) Northwestern University Law Review 803, at p. 812.

simplifying the process of conducting subrogation claims for an MTPL insurer, for instance, with a refutable presumption of a product defect in case that a vehicle caused a crash in automated driving mode would reverse the burden of proof to the manufacturer and could increase the efficiency of subrogation claims. This is because legal costs would only incur in case that OEMs can provide evidence that the accident was caused by the fault of the owner or human driver (e.g. inadequate maintenance, unintended use, …). Consequently, only those accidents, which are verifiably caused by the fault of the owner would be taken to court. As we assume that this will be the minor share of all accidents, legal and processual costs should be reduced.

In addition, we have described how a lack of insurers ability or willingness could lead to an inadequate allocation of liability costs. This is the case if motor insurers, either due to a lack of technical and legal know-how or because of a lack of motivation, fail to prove producer´s liability where the conditions of producer´s liability are actually fulfilled. Here, refutably presuming a product defect could also support that liability costs are allocated based on the factual responsibility for loss causation in a specific loss event.

Is artificial intelligence the solution to all our problems? Exploring the applications of AI for automated driving

Dr. Stefan Milz, Team Lead Algorithm Driver Assistance Research, Expert AI, VALEO

Jörg Schrepfer, Leiter Fahrerassistenz-Forschung Deutschland, VALEO

T. Bertram (Hrsg.), *Automatisiertes Fahren 2019*, Proceedings,
https://doi.org/10.1007/978-3-658-27990-5_10

Abstract

Deep Learning and AI has become the standard model for object detection and recognition such as situation understanding, prediction and planning. In this chapter we explore how AI can be used to improve parts of the classical ADAS algorithm chain. Based on this, we investigate the full Automated Driving pipeline. Firstly, we describe the building blocks of the pipeline composed of standard computer vision tasks. We provide an overview of use cases for automated driving based on the authors' experience in commercial deployment, e.g. Sensor-Fusion, Perception, SLAM or End-2-End Driving. Finally, we discuss the opportunities of using AI and Deep Learning to improve upon state-of-the-art classical methods.

1 Introduction.

Automated Driving (AD) is a rapidly advancing application area with a complex structure (see Fig.1) and lots of progress in Deep Learning. There are two main paradigms in this area:

1. The mediated perception approach which semantically reasons the scene and determines the driving decision based on it.
2. The behavior reflex approach that learns the driving decision End-to-End. The behavior reflex methods can benefit from semantic reasoning of the environment.

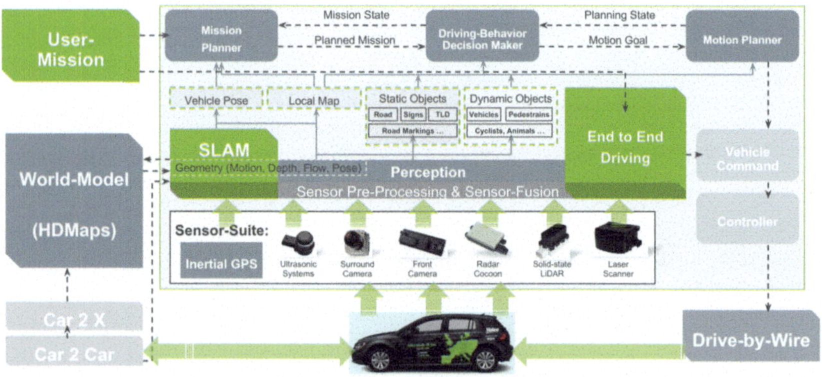

Figure 1: Functional blocks of Automated Driving. AD is composed by several algorithmic blocks which are related to the basic questions of Autonomy: How do I see ? (Sensor Preprocessing and Fusion), Where I am ? (SLAM), Who is around me (Perception)?, Who can I get to my destination (Prediction and Control). Nearly all blocks are influenced by AI. We will outline possibilities.

The structure of an AD system can be roughly divided into the function blocks depicted in Figure 1. The raw data collected by the sensors is processed and evaluated by a series of different algorithms. In this context, it is generally distinguished between classification and localization processes. Classification processes provide information on the type of object and its location, while localization processes determine the actual position of data points which are created by the objects. Combining both enables assumptions on the position, orientation and movement of real objects to be made. The analyzed data from the different sensors is then brought together, so merged, creating a virtual model of the environment including traffic lanes, all types of objects and free space. Possible driving trajectories can be generated and evaluated within this model. Once the system has calculated the lane course, the trajectory can be implemented via control signals sent to the longitudinal and lateral control units. Ego-localization [4] within the environment model is carried out implicitly by the sensor signals and matching them with either a standard ADAS map or a so-called HD map. This form of localization is sufficient for application cases such as those occurring during highway driving. In urban areas, however, the situation becomes more difficult. These scenarios require the mapping of complex traffic-lane scenarios, which is why it is vital to determine the position of the vehicle down to the nearest centimeter. This high localization precision is achieved using the Simultaneous Localization and Mapping (Slam) method. This Slam method use data from sensors like cameras, laser scanner or radar to create an accurate 3D model or map of the vehicle's path as well as to provide the vehicle's precise position and orientation in this map. Redundancy in localization is achieved by additional inertial sensors and GPS feature in the car, while the latter remains lacking availability in many situations.

2 Sensor Preprocessing and Sensor Fusion.

2.1 Sensor Types

A high degree of coverage and redundancy in environmental detection is required for level 3 automated driving functions and above, which allow the driver to take his eyes off the road for a period of time. The sensor equipment therefore usually includes all tried-and-tested technologies, such as ultrasonic, radar and cameras.

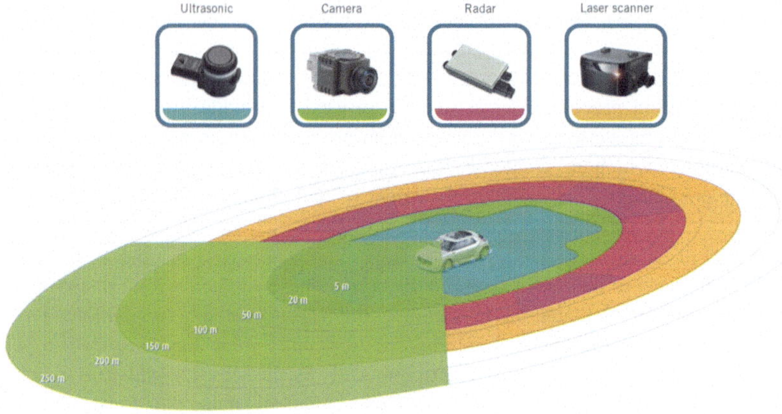

Figure 2: 360 degrees of perception: Different sensors cover different angles and depths around the autonomous vehicle (e.g. Ultrasonic, Cameras, Radar and Lidar)

A new addition is lidar technology. The new Scala laser scanner, Figure 3, developed by Valeo, will be used for the first time in the Audi A8. Thanks to its unique combination of high range and wide perspective paired with great accuracy and high resolution, it is an ideal addition to the available sensor portfolio. The sensors must always keep track of the complete surroundings of the vehicle. In practice, this means multiple overlapping and redundant sensor cocoons. The front sensors (camera, radar, laser scanner) are supplemented by a radar cocoon with four radar sensors in the vehicle corners, a cocoon which uses the four cameras of the surround view system, and the ultrasonic sensors all around the vehicle. The first level 3 systems will be introduced without automatic lane changing function, as this feature requires additional monitoring of the side and rear areas. The addition of a lidar cocoon to supplement the environmental detection is one solution, Figure 2. Valeo has already received an order for this type of cocoon with multiple laser scanners.

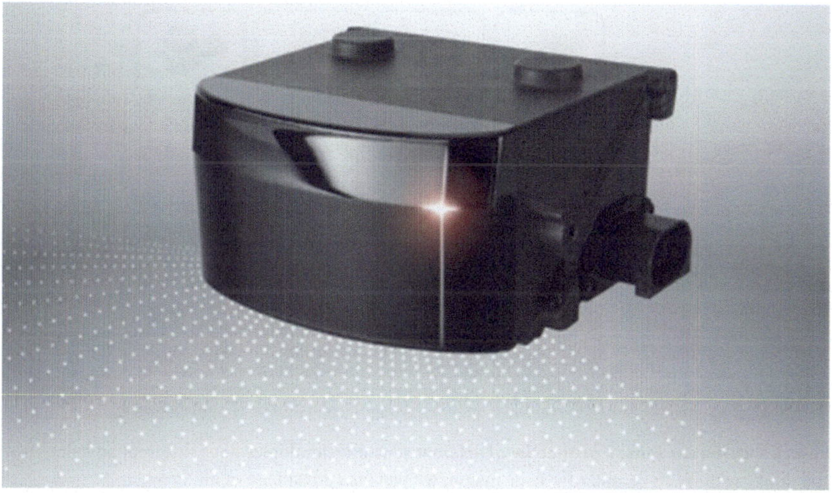

Figure 3 Scala laser scanner with lidar technology for great accuracy and high resolution (© Valeo)

2.2 Sensor Strengths and Power of Sensor-Fusion

Every type of sensor technology has specific physical advantages and disadvantages, which are dependent on aspects such as weather and lighting conditions as well as the driving speed. For a redundant system design, it is vital that not only more sensors are used, but that the sensors used are different and possess characteristics which complement the other sensors. Different technologies are compared to each other in Figure 4. It is clear that laser scanners are a great addition to the other available sensors, and that only sensor fusion, that is a group of sensors, each compensating for the weakness of the other sensors, can lead to the desired result. In order to achieve genuine redundancy in all possible cases, at least two different sensors must provide a usable signal at any given time. This is why three-fold sensor coverage is considered as a requirement for all systems above automation level 2, supplemented by close-range ultrasonic

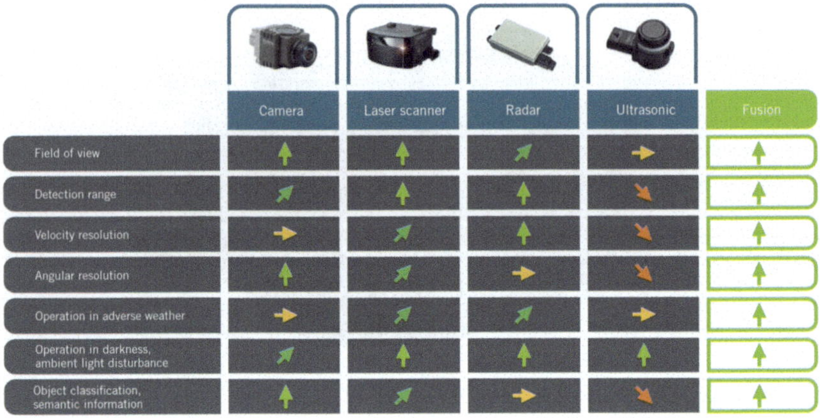

Figure 4 Different sensor properties: Different sensor technologies have different advantages and drawbacks. With the aid of sensor fusion, we can take advantage of all.

2.3 AI based Sensor Fusion Strategies for Automated Driving

In the previous chapter the power of sensor fusion has been pointed out, i.e. Best results either in AI based perception or traditional perception always depend on the quality of the processed data. Hence, best results in scene understanding or planning and control can only be achieved with the best data stream. However, fusion strategies in the field of Deep Learning and AI are different to the traditional domain. However, in the context of AI, the research community for multimodal learning described several methods to perform sensor fusion using convolutional neural networks (CNN). Figure 5 describes the basic idea beyond a CNN for sensor data processing in the AD domain. However, it is more simple to process single sensor data. When it comes to multi-sensor data it is getting more complex. Figure 6 describes six common strategies for the example of pedestrian detection in a 3D space. In this example a fusion of camera images and lidar point clouds is presented. (1-4) describe a sensor fusion a different stages of the used neural network. The network wants to predict 3D Boxes around pedestrians, which is the output. Therefore, two inputs data streams are processed, a camera image and an accompanied Lidar point cloud. Both are fused inside the network. Another fusion approach is a consecutive processing (5). A prefiltering of the point cloud is done using a camera image. Last presented fusion strategy describes fusion over the time domain.

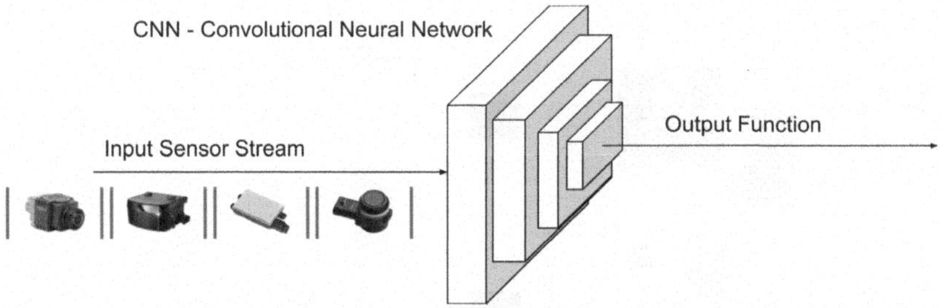

Figure 5 CNN for Automated Driving: A very common Deep Learning model is a Convolutional Neural Network, because it is very suitable for sensor data processing, e.g. cameras. It can be used for any kind of function, like object detection, semantic segmentation, environmental modelling or planning and control

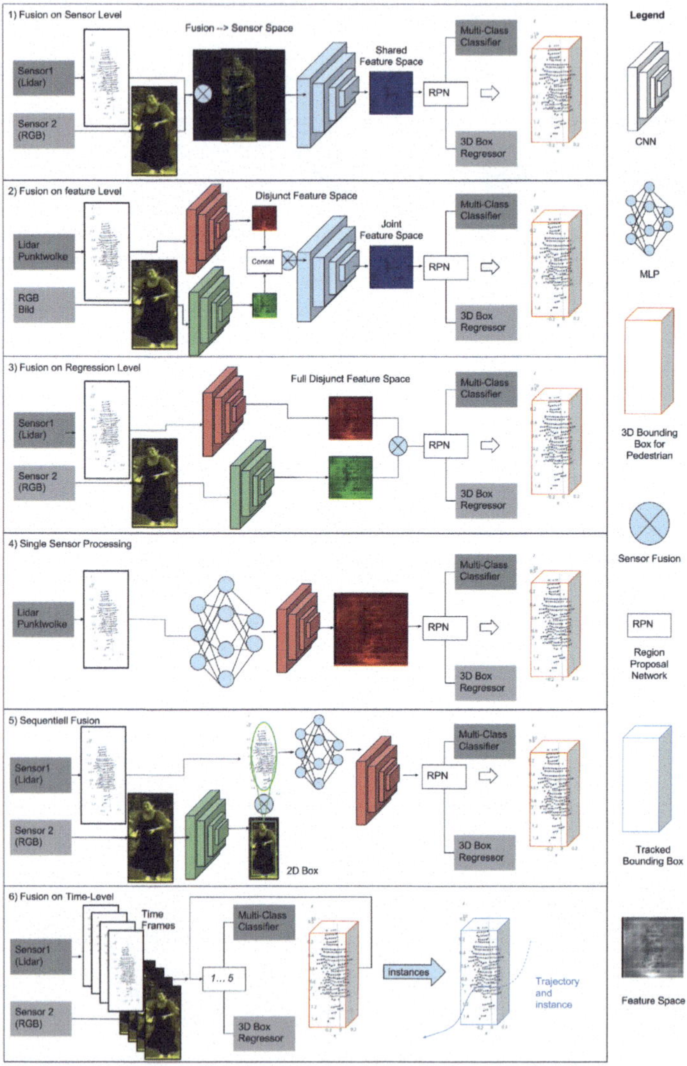

Figure 6 Multi-Sensor Fusion strategies using CNNs : Different sensor technologies have different advantages and drawbacks. With the aid of sensor fusion, we can take advantage of all. The figure outlines six AI based sensor fusion strategies for the example of 3D pedestrian detection using lidar and cameras: e.g. early fusion (1), mid-fusion (2), late fusion (3), single processing (4), sequentiell fusion (5) and tracking.

3 Perception

Figure 1 outlines the functional blocks of the automated driving domain. A critical part is the environmental perception. The aim is to detect any kind of static and dynamic objects around the vehicle. Since, the perceptual analysis is the base for further planning and control, a very accurate and real-time capable analysis is needed. Due to this critical requirements, recent leading methods are all based on AI especially Deep Learning. The following investigation describes three case studies for environmental perception (camera, lidar, ultrasonic), all based on Valeo's sensor technology.

3.1 Camera based 360 degree Semantic Segmentation

Visual semantic segmentation is a pixel-wise classification of camera images. Each pixel belongs to a class, predicted by specifically trained CNNs. The CNN predicts classes with confidences and is therefore enforced to output on class (highest confidence) per pixel. Class types refer to the functional approach, e.g. cars, pedestrians, road or vegetation. Figure 7 emphasizes a specific model for Automated Driving that interprets the total environment of the vehicle using Valeo's fisheye cameras. The fisheye cameras have a very large field of view (> 180 degrees) in horizontal direction. With the aid of those specific lenses a full perception could be achieved using four fisheye cameras. Hence, four images are individually inspected by the AI and the stitched for full semantic perception, Figure 7.

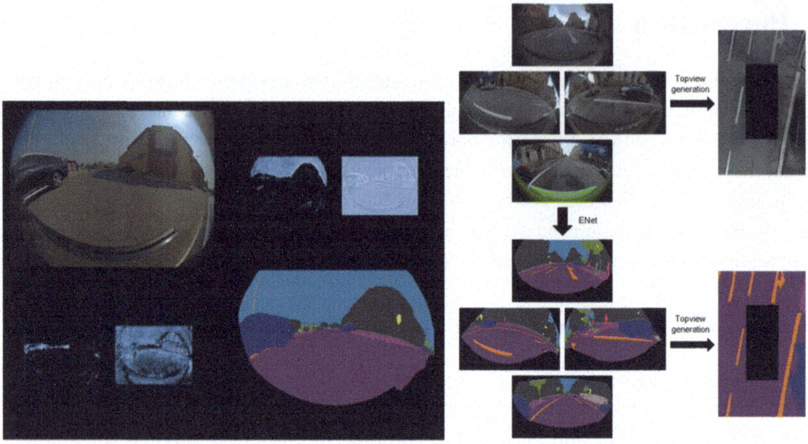

Figure 7 Visual Semantic Segmentation on four Valeo Fisheye Cameras: For a seamless 360 degree semantic perception up to 20 meters four single fisheye cameras are inspected individually (upper right). The single semantic segmentation results are stitched targeting a composed environmental model (lower right). The left part shows what a CNN (i.e. the ENet model in this case) "sees" within the camera image to perform pixel-wise classification on the fisheye cameras (e.g. purple → freespace)

3.2 Lidar based 3D Box Detection

Another important task within the perceptual block, is the 3D multi-class bounding box detection. For planning and control strategies is it important to know which dynamic objects behave in what manner around the self-driving car. Therefore, it has been established to represent those objects as boxes, that could be tracked and classified for further predictions. It is crucial to percept a precise location and orientation of those objects. To cover a huge range with a precise depth, it is necessary to use a Lidar sensor. Recently, Valeo presented one of the fastest and precise Lidar based 3D bounding box detectors on Lidar, Complex YOLO, Figure 8.

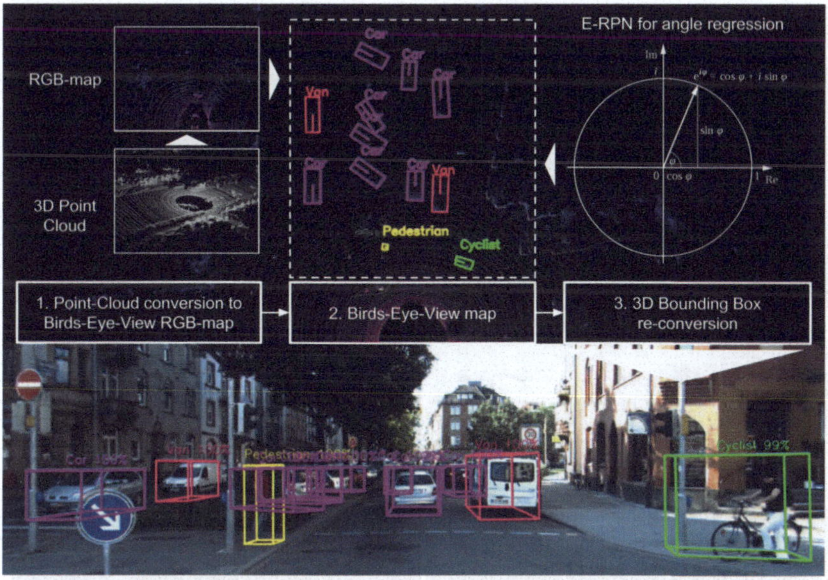

Figure 8 Lidar based 3D Box detection using Valeo's complex YOLO: For a very fast and accurate 3D Bounding Box prediction, the lidar point cloud of the vehicle is converted into an RGB-Birdseye-View Map (RGB). This map is processed with the Complex YOLO, which is trained to detect multiple (8 classes) boxes in 3D (x,y,z and yaw angle). The approach is lidar based only. This figure shows 3D boxes in camera space only for visualization.

3D Boxes have to be used within the environmental model. On top a box tracking is necessary to incorporate tempo-spatial features. The method can also be combined with semantic segmentation. Figure 9 outlines a prestage of an environmental model, covering 3D object detection and tracking, such as visual fusion using semantic segmentation.

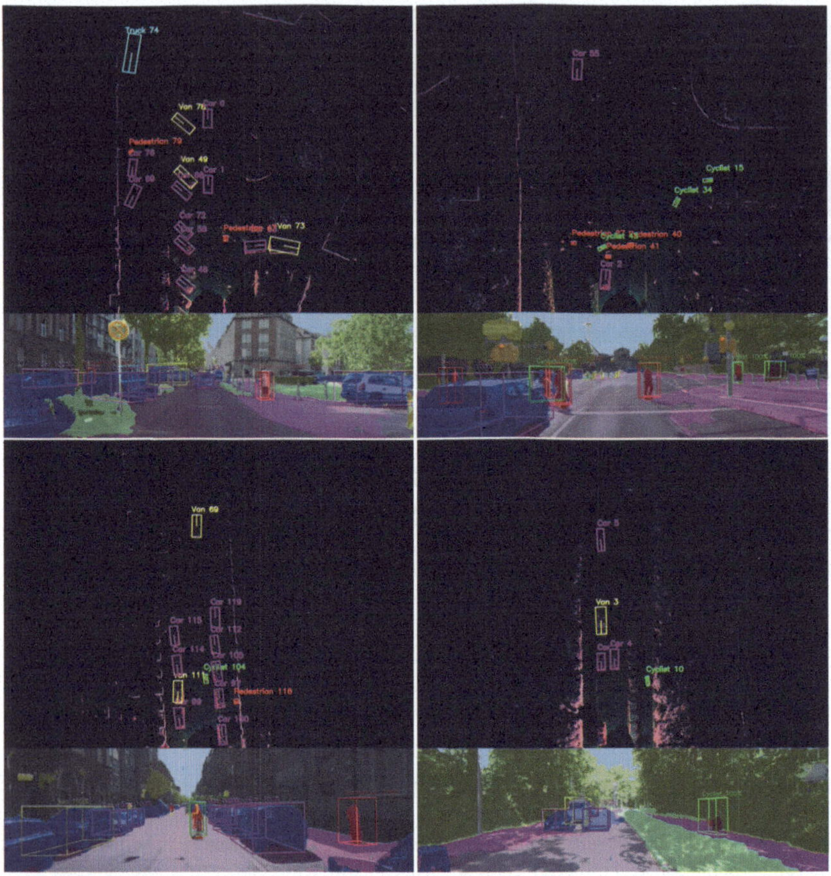

Figure 9 Environmental model prestage: This figure shows qualitative results of the Complex YOLO based 3D Box detection and tracking such as the combination with visual semantic segmentation (ENet, Figure 7). This results in an instanced based environment pre-model with full pixel wise interpretation and tracking for dynamic objects

3.3 Ultrasound based height Classification

A very underestimated sensor technology is the ultrasonic sensor. However, it is very trustworthy for range estimation ibn the absolute near range. Due to its low price, it is the most common sensor for parking scenarios. In the past, the technology was less used for object classification or height estimation. With the aid of CNNs, especially Capsule based neural networks, Valeo presented recently a novel method to classify heights of objects using a single Valeo ultrasonic parking sensor. For this procedure, the raw sensor signal is transformed into a complex RGB map, i.e. the R channel inherits the absolute, the G channel complex part of the signal, see Figure 10. The map was feeded into the network. The model is trained to detect four different classes, that are crucial for parking. The outcome was an accuracy of more than 95% in the test-set.

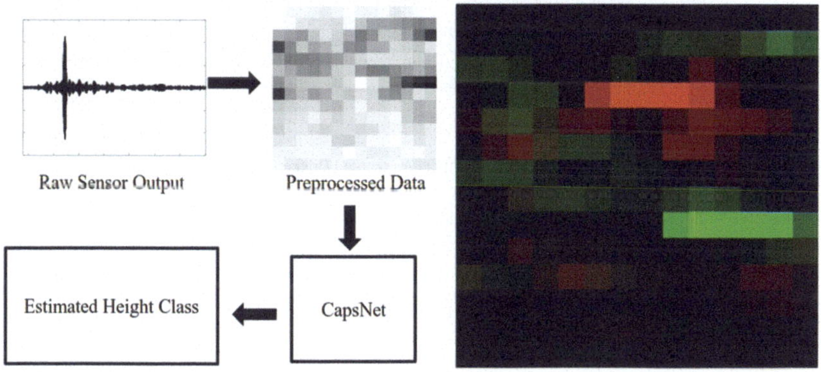

Figure 10 Ultrasonic height estimation: This figure shows the processing pipeline of ultrasonic based height estimation. The raw sensor signal is converted into a complex map (right: R channel → absolute, the G channel complex part of the signal) which is processed by a Capsule Network (4 classes, 95% accuracy)

4 SLAM

Another functional block (Figure 1) and not less important part is the localization of the vehicle. Within the robotics domain, localization is always interlocked with the mapping itself and therefore called SLAM (simultaneous localization and mapping). The field of SLAM has a strong history, especially for Vision, and could be itself separated into functional blocks, where almost all are influenced by AI techniques recently, Figure 11. This chapter presents single case studies within the general SLAM pipeline solved using AI.

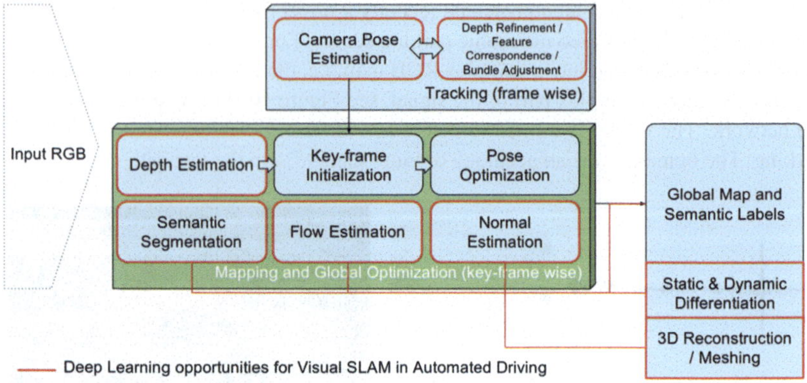

Figure 11 Visual SLAM Pipeline. This figure investigates the opportunities for AI and Deep Learning within the visual SLAM pipeline. E.g. Visual Depth Estimation and Camera pose estimation are recently solved using Deep LEarning models.

4.1 Visual Depth Estimation

Figure 11, promotes monocular visual depth estimation as very important task for solving SLAM. Recently almost all leading models are based on Deep Learning. In comparison to visual semantic segmentation (Figure 7), this is not performed as pixel-wise classification, but as pixel-wise regression. The specific model tries to predict a accurate depth per pixel. Figure 12 shows qualitative results of Valeo's unsupervised approach. This method is trained on stereo camera images, but can perform on monocular cameras during inference (runtime)

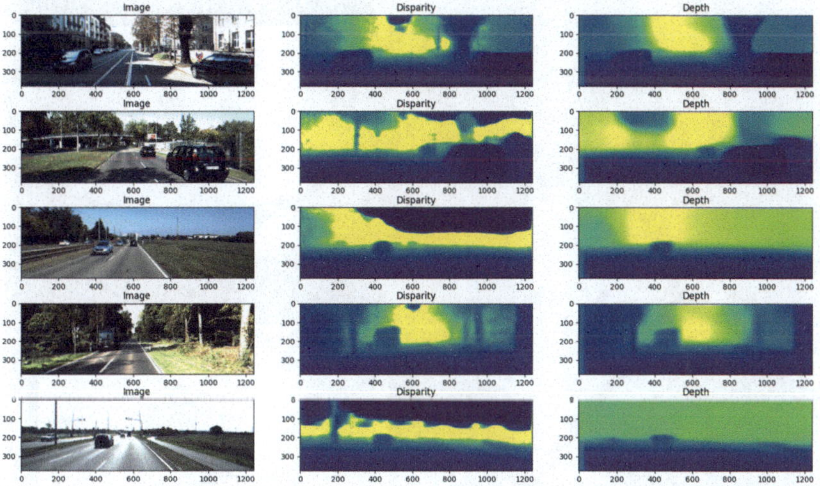

Figure 12 Visual Monocular Depth Estimation. This figure shows results for monocular visual depth estimation using an unsupervised approach. The left column shows input images, the middle column presents depths by predicting disparities, the right column shows direct depth predictions.

4.2 Visual Odometry and Mapping

A precise depth estimation could be used for further camera pose estimation and a simultaneous mapping. Again, AI methods are used for relocalization. Figure 13 shows a final odometry graph based on Valeo's fisheye front camera. The visual Depth estimation is therefore used for mapping. The results show the visual mapping of beautiful city Kronach (Bavaria). However, since depths are always predicted and not directly measured, those results still lack in accuracy compared to lidar based SLAM methods.

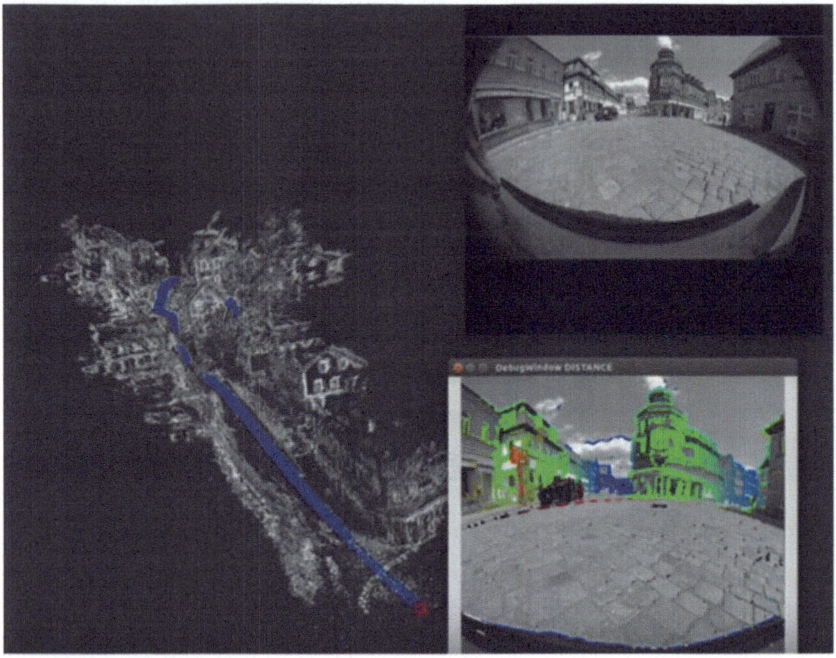

Figure 13 Fisheye based Visual SLAM. This figure shows results for monocular visual SLAM using Fisheye cameras. The right upper part shows the input image, the right lower part shows a sparse depth estimation. The graph inthe left presents the final map with a blue odometry trajectory provided by the SLAM localization.

4.3 Semantic Lidar based SLAM

Lidar based SLAM techniques outperform Visual SLAM approaches by far in terms of odometry and mapping quality. Due the fundamentalism of Lidar (Figure 4), depths could be measured directly and therefore only the registration task must be performed. However, Lidar is much more expensive and therefore cameras can be the better choice. A famous approach for Lidar SLAM is LOAM (Lidar Odometry and Mapping). Figure 14 presents a combination of LOAM and Visual Semantic Segmentation (Figure 7). This leads to mapped point cloud with semantic metadata as a perfect base for planning and control in the domain of Automated Driving. The semantic interpretation helps the SLAM to be more stable due to more robust feature matching.

5 Environmental Model and Prediction.

A very late block within the Automated Driving pipeline is the the environmental model, which is the base for the prediction and planning such as the control part. In section 4, the semantic based SLAM, in section 3, the semantic 3D box tracking were presented. Both together end up in a environmental model, which is discussed in this section. On Top, we end up in a prediction chapter, mainly based on Waymos ChaffeurNet proposal.

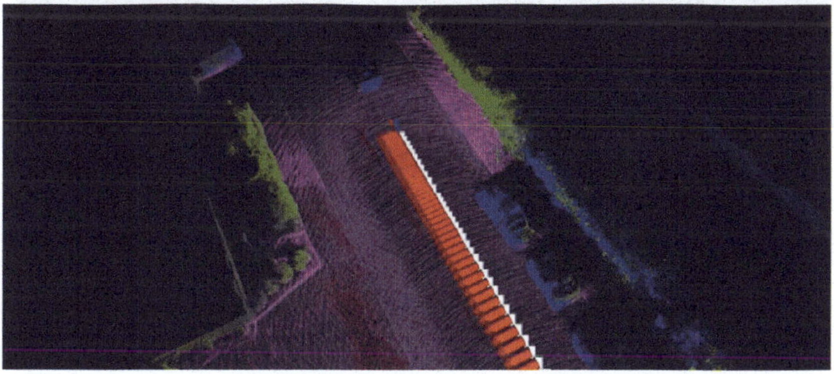

Figure 14 Visual Lidar based SLAM. This figures shows a mapping and odometry result by using LOAM and a visual semantic segmentation. The red trajectory shows the estimated path of the vehicle. The surrounding pointcloud is semantically interpreted by AI: e.g. purple points belong to the street, green points belong to vegetation and blue points belong to static cars.

5.1 Combined Environmental Model

The combined environmental model is the base for planning. Therefore, it requires to have the vehicle pose, the full perception of the surroundings, the full mapping of the surroundings, a 3D box estimation and instantiation (tracking) of static objects. With the aid of the previous AI solutions its possible to combine all together.

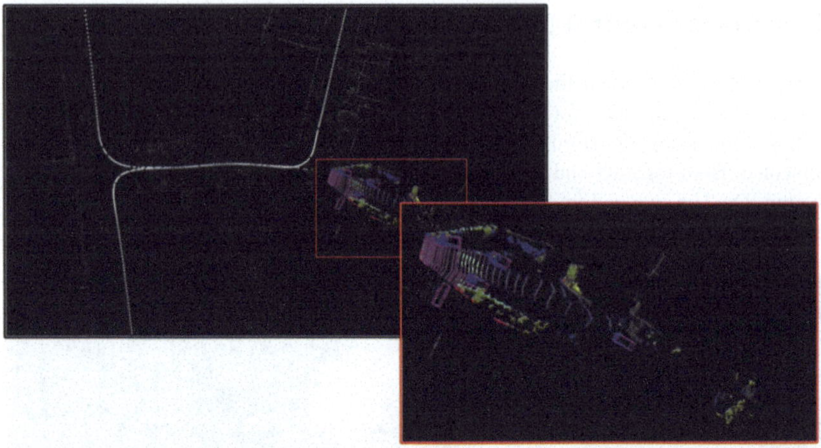

Figure 15 Environmental Model. The figure outlines an environmental model containing Map (left), vehicle pose and track (white pose), perception (semantic segmentation right) and 3D Boxes such as their instance tracks.

5.2 Prediction

Once environmental instances and maps are provided, prediction analysis could be performed. Figure 16, shows qualitative results provided by ChauffeurNET (Bansal et al.). All the model predictions are based on CNN regressions using the environmental model as an input. Therefore an abstraction of the objects is done. Different abstraction layers are feeded into the network to predict the future behavior of the objects.

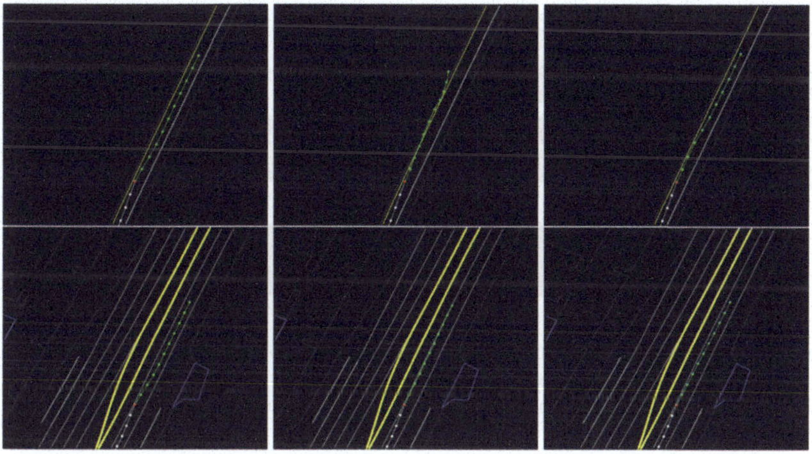

Figure 16 Prediction by ChauffeurNET. This Figure is taken from Waymos Chauffeur-Net (Bansal et al.) and presents qualitative results of object trajectory predictions. White and yellow lines show static objects like lanes etc. The Green outlines the future pose and the white the past pose. The left column describes the ground truth and the middle and right two different predictions.

6 End-2-End Driving

We have seen seen that almost all functional blocks are meanly leaded by AI based solutions. Another important research area is the End-2-End driving domain (Figure 1, Figure 17). Valeo has recently shown that End-2-End driving is possible using CNN networks. An advantage is here, that it can be performed using only cameras. The network tries to imitate the human behavior. Therefore, it directly learns steering and acceleration such as de-acceleration of the car. The training data set is simply extracted by observing human drivers, where cameras streams and bus informations is recorded. Results are impressive, but the solution has a important drawback, because it's mainly impossible to debug and to validate the performance of the network. Hence, the idea is to use it as a redundant path within the Automated Driving pipeline. It is used as a fallback solution for corner cases situations (see Figure 17).

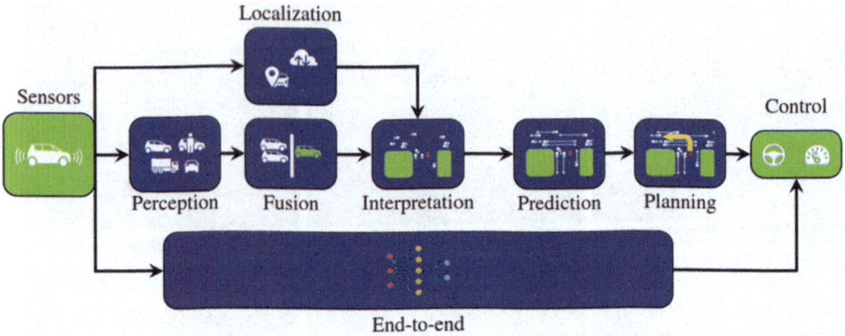

Figure 17 Valeo's End-2-End Driving. The End-2-End driving approach is used as parallel path within the Automated Driving pipeline. The network directly learns to imitate the human behavior (e.g. steering). It is a good fallback solution for corner case situations and dead-locks.

7 Conclusion

Within this chapter we have presented a strong overview of the Automated Driving pipeline. We have seen that in almost all functional blocks, AI solutions do outperform traditional methods. However, due to the fundamentalism of CNNs and AI, conventional approaches must be re-considered (e.g. Sensor Fusion, Figure 6). AI based solutions are blessing for the complex situations. However, despite all the outstanding performance, we should not forget about the basic strengths of different sensor technologies. Hence, an over engineered AI model based on a weak sensor should not be a solution in critical area like Automated Driving.

REFERENCES

[1] Valeo: Automated driving challenges with Valeo Cruise4U. Online: https://www.youtube.com/ watch?v=MJWrVyL1_n8, access: 27 October 2017

[2] Navya: Navya launches autonom cab, the first 100-% autonomous robo taxi. Online: http://navya. tech/en/press/navya-launchs-autonom-cab-inparis/, access: 22 November 2017

[3] Schaller, T.; Uhler, W.; Schrepfer, J.: Proposal for standardization of sensor interfaces to a fusion unit"; VDI conference Eliv 2017, Bonn, 18 and 19 October 2017

[4] Jo, K.; Bradai, B.; Resende, P.; Picron, V.: Precise ego-localization using multi sensor map matching for highly and fully automated vehicles. SIA Cesa 2016 Congress, Paris, 16 and 17 November 2016

[5] Spandfelder, B.: Challenges in applying the ISO 26262 for driver assistance systems. 5th Conference on Driving Assistance, Munich, 15 and 16 May 2012

[6] ISO/TC 22/SC 32: N0356 Standardization of sensor interfaces. German proposal, ISO SC 32 sensing workshop, Berlin, 6 and 7 March 2017

[7] Fitzek, F.; et al.: Cross-industry cooperation for camera technology standardization. VDI conference Eliv 2017, Bonn, 18 and 19 October 2017

[8] Reilhac, P.; Moizard, J.; Kaiser, F.: Cockpit Concept for Conditional Automated Driving. In: ATZworldwide 118 (2016), No. 3, pp. 42-46

[9] DLR Cologne: Pegasus Research Project. Online: http://www.pegasus-projekt.info/en/home, access: 27 October 2017

[10] Bradai, B.; Picron, V.; Heitzmann, T.; Maurel, P.; Randriazanamparany, R.; Taccoen, D.; Slimi, H.: Automated driving for all: entry mated. SIA Cesa 2016 Congress, Paris, 16 and 17 November 2016

[11] Valeo: Valeo launches Valeo.AI, the first global research center in artificial intelligence and deep learning for automotive applications based in Paris. Online: http://www.valeo.com/en/valeo-launchesvaleo-ai-the-first-global-research-center-inartificial-intelligence-and-deep-learning-for-automotive-applications-based-in-paris/, access: 27 October 2017

[12] Simon,S et al: Complex-YOLO: Real-time 3D Object Detection on Point Clouds

[13] Bansal et al: ChauffeurNet: Learning to Drive by Imitating the Best and Synthesizing the Worst

[14] J. Schrepfer, J. Mathes, V. Picron, H. Barth: Automated driving and its sensors in test. ATZ Journal January 2018

Artificial intelligence for automated driving – quo vadis?

Dr. Alexander Jungmann

Dr. Christian Lang

Dr. Florian Pinsker

Dr. Roland Kallweit

Mirko Taubenreuther

Dr. Matthias Butenuth

T. Bertram (Hrsg.), *Automatisiertes Fahren 2019*, Proceedings,
https://doi.org/10.1007/978-3-658-27990-5_11

1 Introduction

Making Automated Driving (AD) a reality is a challenging task and still subject of intensive research and development. The core component for realizing AD beside the actual vehicle platform and the sensors for perceiving the environment is the AD system, i.e., the software that enables a vehicle to perform distinct dynamic driving tasks in a distinct operational design domain. Briefly speaking, an operational design domain defines environmental and time-of-delay conditions and restrictions, respectively, the presence or absence of certain traffic, and roadway characteristics. A dynamic driving task, in turn, refers to the actions to be performed, such as lane keeping, overtaking, stopping at a traffic light, or parking in a parking slot.

In order to increase the level of driving automation of a single vehicle, AD systems have to be extended in two dimensions. On the one hand, a continuous automation without relying on human drivers can only be achieved by making vehicles more intelligent, i.e., by implementing high-level decision-making techniques that enable a vehicle to plan and perform situation-dependent driving maneuvers. In addition to the level of driving automation, the scope of the respective operational design domain an AD system is approved for has to be continuously expanded. That is, driving functions have to become more intelligent to be able to handle situations that are more complex.

In this work, we discuss how Artificial Intelligence (AI) already facilitates the realization of more intelligent AD systems. Furthermore, we try to look into the future of AI for AD and identify what is yet to come. Of course, this is a rather subjective view and cannot be considered completely unbiased.

2 What is Artificial Intelligence?

Before taking a closer look at the past, the present, and the future of AI, let us briefly clarify what AI actually means by sticking to some common definitions composed by [1]. The four definitions shown in Table 1 are organized with respect to two dimensions:

1. Reasoning (top) vs. behavior (bottom):

   Reasoning approaches deal with the actual thinking. Within the context of AD, such deliberative approaches enable vehicles to plan imminent maneuvers or even a complete sequence of subsequent maneuvers for anticipatory driving. Behavior, in turn, is more related to the concrete acting process. That is, behavior-based approaches are responsible for transferring high-level decisions into reality [2].

Table 1: Several definitions of AI [1].

Thinking Humanly	Thinking Rationally
"[The automation of] activities that we associate with human thinking, activities such as decision-making, problem solving, learning…" [3]	"The study of the computations that make it possible to perceive, reason, and act." [4]
Acting Humanly "The study of how to make computers do things at which, at the moment, people are better." [5]	**Acting Rationally** "Computational Intelligence is the study of the design of intelligent agents." [6]

2. Human performance (left) vs. rational performance (right):

   The major difference between both approaches is whether success is measured in terms of fidelity to human performance or against an ideal performance. For example, from a low-level control algorithm that is responsible for lateral and longitudinal control of a vehicle, we first and foremost expect a rational performance, i.e., the algorithm should always do the right thing, given what input data is available. For a planning algorithm, which is responsible for high-level decision-making, a performance that is similar to human decision-making is usually more appropriate.

The previous definitions are indeed still quite abstract and do not give much indication about concrete AI research areas, especially not with respect to AD. We will provide concrete examples in the further course.

3 The Past: History of AI

In order to better understand and evaluate current and upcoming trends in the field of AI, let us briefly cover how AI evolved from the very beginning in 1943. For more information, please refer to [1], from which we derived the following key facts.

1943-1955: AI is not yet born

The first work that can be recognized as AI appeared. It was inspired by the knowledge of the neurons of human brains. The first neural network computer was built by using 3000 vacuum tubes and a surplus automatic pilot mechanism from a B-24 bomber to simulate a network of 40 neurons.

1956: Birth of AI

The first computer program that was able to think symbolically instead of numerically was constructed. Furthermore, AI became a separate field and emancipated itself from fields such as control theory and decision theory. AI embraced the idea of duplicating human skills while simultaneously being a branch of computer science. Furthermore, AI was the only field to attempt to build machines that will function autonomously in complex, dynamic environments.

1952-1969: Great expectations

The program "General Problem Solver" was the first program to imitate human problem-solving protocols and could handle a limited class of puzzles. Several other programs for checkers or draughts were developed. These programs eventually learned to play at a strong amateur level. In addition, the program "Advice Taker", which can be considered as the first complete AI system, could achieve competence in new areas without being reprogrammed.

1966-1973: Reality strikes

Earlier systems turned out to fail when tried out on problems that are more complex. Most early programs knew nothing of their subject matter, but solving complex problems requires background knowledge, leading to the famous mistranslation "the spirit is willing but the flesh is weak" to "the vodka is good but the meat is rotten". Intractability of many of the more complex problems could be handled neither by existing problem-solving programs nor by simply applying faster hardware or larger memories. The basic structures (such as simple neural networks) that were previously used to generate intelligent behavior were too limited. Ironically, the backpropagation algorithm for multi-layer networks, which is widely used today, was already discovered.

1969-1979: Knowledge-based systems

Expert systems containing large numbers of special-purpose rules (domain-specific knowledge) were able to solve problems that are more complex. A large number of different representation and reasoning languages were developed, while a clean separation of knowledge (in terms of rules) and reasoning component became important.

1980-present: AI becomes an industry

The first commercial expert system began operation. By 1986, it was saving the company an estimated $40 million a year. More companies followed over time. Overall, the AI industry boomed from a few million dollars in 1980 to billions of dollars in 1988, based on expert systems, vision systems, robots, and software and hardware specialized for these purposes. Unfortunately, the so-called "AI Winter" followed. Many companies fell by the wayside because they failed to keep their ambitious promises.

1986-present: Return of neural networks

The backpropagation learning algorithm was reinvented, leading to so-called connectionist models of intelligent systems. The respective results caused great excitement.

Originally, connectionist models were considered as direct competitor to the symbolic models used in expert systems. Today, both approaches are considered complementary.

1995-present: Emergence of intelligent agents

Researchers intensified the research on complete agents. The results of previously isolated subfields of AI had to be tied together. In particular, sensory systems cannot deliver perfectly reliable information about the environment. Consequently, reasoning and planning systems must be able to handle uncertainty. Furthermore, AI has been drawn into much closer contact with other fields. For example, progress in the control of robotic cars has been derived from better sensors, control-theoretic integration of sensing, localization and mapping, as well as a degree of high-level planning.

2001-present: Very large data sets

So far, the emphasis has been on the algorithm as the main subject of study. However, new work suggested, that for many problems, it makes more sense to worry about the data and to be less meticulous about what algorithm to apply. The amazing results of Deep Learning (DL) in particular fields such as image processing, where multi-layer neural networks are trained with large data sets to detect objects, confirmed this.

Final remarks

Machine Learning (ML) is in fact a subfield among many others of AI. Due to the increasing performance of hardware and the increasing amount of available training data, however, we are confident to say that ML is currently the most dominant and propagated subfield in the AI research. Results in different domains already demonstrated the potential of ML techniques, while handcrafted algorithms are often outperformed.

Machine Learning is usually separated into three areas: Supervised Learning based on labeled training data (e.g., for classification tasks), Unsupervised Learning without any teaching entity (e.g., for clustering tasks), and Reinforcement Learning (RL) [7] for learning by interacting with the environment, e.g., for appropriately selecting actions given a specific situation.

4 The Present: AI for Automated Driving

Given the definition as well as a brief overview of the history of AI, we will now connect AD with AI. After introducing basic components of an AD system, we introduce how AI is already supporting AD in general, while also giving some concrete examples.

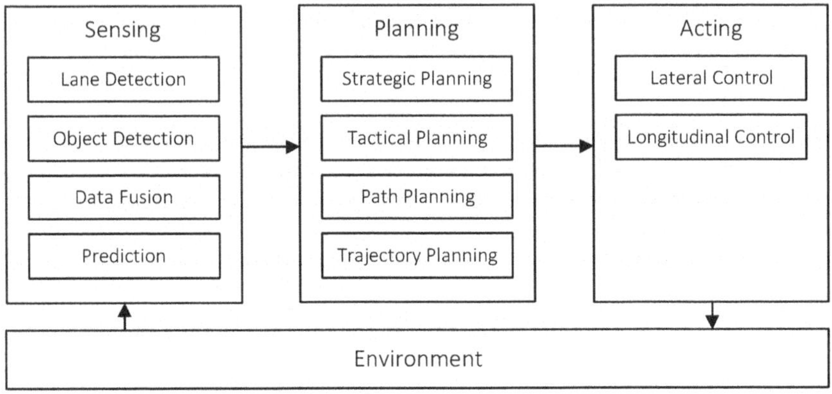

Figure 1: Overview of a prototypical AD system.

Automated Driving System

Figure 1 shows some basic functional components of a typical AD system. It also illustrates the interaction between these components and between the components and the environment, resulting in a closed control loop. Please keep in mind, that the depicted components are just an excerpt of a significantly more complex system. Functional components can usually be clustered into sensing, planning, and acting. The sensing cluster (also referred to as perception; we use both terms synonymously) is responsible for perceiving and understanding the environment based on vehicle sensors such as cameras, LiDAR (light detection and ranging device) sensors and Radar sensors. In the sensing cluster, we typically find algorithms for

- extracting and fusing lane information from different sources,
- detecting, classifying, tracking, and fusing dynamic objects (e.g., other traffic participants) as well as static objects (e.g., parked vehicles, traffic lights and signs), and
- predicting the behavior of dynamic objects.

The planning cluster is responsible for implementing the deliberative behavior of a vehicle. Based on the perceived environment and derived situation, the planning cluster chooses the actual course of action and can therefore be considered as the main intelligence or brain of the AD system. Typically, the planning cluster is divided into different decision-making and planning levels:

- strategic planning is responsible for high-level decision-making tasks such as dynamic navigation and anticipatory driving while taking different – sometimes even conflicting – goals (e.g., comfort, energy efficiency, or traffic flow) into account,

- tactical planning is responsible for choosing an appropriate maneuver given a concrete environmental situation and according to the decisions made by the strategic planning component,
- motion planning is responsible for computing the actual path (geometry information) according to the previously selected maneuver as well as deriving the distinct trajectory (geometry information + velocity information).

Based on the output of the planning cluster, the acting cluster is finally responsible for generating control variables such as acceleration and steering angle.

Isn't this already AI? And what about Machine Learning?

Sensing (e.g., by means of computer vision techniques), planning, reasoning etc., which are core components of an AD system, are in fact classical fields of AI. Even the overall sensing-planning-acting approach is a typical approach from robotics, which is in turn a classical field of AI. In specific areas, however, ML techniques are dominating classical algorithms by this time. DL techniques, e.g., outperform classical computer vision algorithms in terms of detection and classification rate, even when the complexity of the environment increases. ML techniques also excel in predicting the behavior of other traffic participants as well as inferring additional information that is initially hidden in the data produced by the detection and classification processes.

As already mentioned, decision-making within the planning cluster is fundamental in order to realize autonomous vehicles in the end. With an increasing scope of the corresponding operational design domain, we need solutions that allow for decision-making and planning with uncertainty and can even adapt to unforeseen situations. Classical approaches such as handcrafted rule-based systems may work for a very restricted and well-defined operational design domain in combination with a limited set of dynamic driving tasks to be performed. Nevertheless, they will soon hit the wall. It is almost impossible (but most certainly extremely time-consuming) for a developer to foresee during the development process all possible situations a vehicle may face. A well-known mathematical framework for planning with uncertainty are Markov Decision Processes (MDPs) from the AI domain. An MDP can inherently handle uncertainty in a sequential decision-making process. Furthermore, MDPs are the mathematical basis for RL. Instead of manually defining which maneuver shall be performed in a certain situation, an RL-based approach learns how to make this decision by interaction with the environment, e.g., by imitating a human driver or by learning completely from scratch in a simulation environment.

A typical strategy for improving a pure sensing-planning-acting system is to incorporate additional ML algorithms, to enhance existing classical AI algorithms by ML techniques or to replace classical AI algorithms by ML algorithms, as it is currently happening in the computer vision domain. As a result, the performance range of the corresponding AD system is gradually increasing. For example, a lane following functionality, which was developed for highway environments, might work in urban environments after improving lane detection and fusion algorithms by means of ML techniques. This is easily possible when the AD system is designed as a white box system being decomposed into many loosely coupled functional modules.

An opposite approach is end-to-end learning. The entire AD system is considered a black box without any pre-defined intelligence in the first place. In fact, Deep Reinforcement Learning (DRL) techniques – a combination of DL and RL – replace the sensing, planning, and acting components, or at least major parts of them. Roughly speaking, in DRL, the underlying decision-making model is not defined in advance, but is replaced by a deep (multi-layer) neural network architecture, which is trained based on RL techniques while interacting with the environment.

We should nevertheless point out that ML for AD is still subject of intensive research. Regarding the sensing cluster of an AD system, however, we are confident to say that ML techniques are indeed well established.

Machine Learning for Automated Driving: State of the Art

Let us now consider a few concrete examples of ML for AD. Those who are interested in a more comprehensive overview might want to have a look at existing surveys that give a general overview of the recent developments in the realm of autonomous vehicle software [8] and provide information about environment perception modelling, object detection, motion planning techniques, motion control, sensors' features, and even take safety and security into account [9].

Sensing

A lot of ML research that is conducted is related to sensing. Camera-based approaches, e.g., use Convolutional Neural Networks (CNNs) – a class of deep neural networks that gained significant acceptance over the last years – in order to detect and estimate the pose of 3D objects from single images [10] [11] [12]. Although the details of the approaches may differ, the basis in terms of CNNs is often quite similar.

The application of ML techniques for sensing is not restricted to camera-based approaches. Neural network-based approaches for 3D object detection [13] [14] or 3D classification and segmentation tasks [15] based on LiDAR data eventually emerged. Camera-based and LiDAR-based approaches can even be combined, e.g., for object detection and classification based on CNNs [16] [17], real-time object recognition [18], or 3D object detection from camera data with depth information [19]. A first comprehensive survey on 3D object detection methods for Automated Driving was recently published [20]. It focuses on end-to-end pipelines and learning techniques. In this particular object detection context, end-to-end learning refers to black box approaches that usually accept sensor data as input and provide object data as output.

Some object detection approaches are more specialized. For example, CNN-based approaches that specialize on vehicle detection are able to improve detection during night scene and in bad weather [21]. Furthermore, camera and LiDAR data can be fused to detect and classify pedestrians more robustly [22] [23]. Even predicting the behavior of pedestrians in urban scenarios [24] or predicting human driving decisions [25] becomes possible. Other approaches, in turn, specialize on sensor-based detection of infrastructure objects such as traffic lights [26] or lanes [27].

Planning and Decision-Making

In comparison to ML techniques for sensing, ML techniques for planning and decision-making are still in the very early stages of development – although the theory already exists for decades. An overview of recent advances in planning and decision-making for AD is given in [28]. Many proposed approaches break open the clean separation of sensing and planning by realizing an end-to-end learning approach. For example, in order to learn to drive in a simplified environment containing only lane markings and static objects, a deep Q-Learning network can be used, with images from the front camera being the input and discrete actions being the output [29]. Deep Q-Learning is a concrete manifestation of DRL. It combines techniques from DL with Q-Learning, which is a well-established RL algorithm. Deep Q-Learning can even be adopted to facilitate continuous control of a vehicle instead of working with discrete actions [30] or to realize human-like vehicle speed control [31]. However, as you can see, even the clear separation between planning and acting becomes blurred. In fact, many of the recently published learning approaches, e.g., for overtaking maneuvers [32], lane change behavior for diverse and even unforeseen scenarios [33], high-precision lane keeping [34], distinct behavior in a traffic light passing scenario, intelligent collision-avoidance [35], or trajectory planning for time varying environments [36], can hardly be classified into the rather classical sensing-planning-acting system.

Machine Learning for Automated Driving at IAV

At IAV, we investigate the utilization of ML for various aspects of AD. Some examples of these activities are briefly introduced below.

We studied the visual detection of drivable road area around the vehicle using different image processing and machine learning approaches. It generally shows that recent semantic segmentation approaches based on deep encoder/decoder neuronal networks perform very well in terms of road segmentation accuracy if the encountered kind of road scene is sufficiently similar to large portions of the training data. In this case, they are superior to approaches based on classical image processing and machine learning techniques. However, they can fail for new, dissimilar scenes that are not covered well by the training data (e.g., some rural scenes). In such situations, classical machine learning techniques applied on manually designed texture features, combined with intelligent heuristics for both the online verification of the segmentation result and the dynamic adaption to new road types can deliver better results.

We also investigate the recognition of road conditions using front camera images. This concerns two important aspects: weather conditions and structural road surface conditions. For the first aspect, we classify images into weather categories such as sunny, cloudy, rainy and snowy, which can have a significant influence on the road surface grip and visibility conditions. To address the second aspect, we classify the drivable road surface in front of the vehicle into structural categories according their expected influence on the required driving behavior of the vehicle (e.g., good asphalt road, cobblestone, potholes, field road, etc.). For the recognition of both kinds of road conditions, we studied various deep neuronal network architectures and state-of-the-art training techniques. It turned out that classification accuracies of 85%-99% can be achieved using current state-of-the-art network architectures for normal, typical situations, which, however, drop significantly for less common or very difficult scenarios.

In order to gain a level of "self-awareness" about LiDAR sensor data comparable to that of a human being, i.e., in order to enable the AD system to detect and realize, when something is out of the ordinary, we trained an artificial neural network to learn when to decide whether the data is compromised or not. Furthermore, we let the neural network classify the degree of corruption of the considered data. In order to generate data associated with compromised vision we partially covered the LiDAR sensor, while data for flawless vision was simply collected with a clean cover of the sensor. We were able to show that a LiDAR sensor can have a primitive sense of self-awareness in the simple way that it realizes whether it is working properly or not.

In another line of research, we examine the prediction of lane change maneuvers of other traffic participants. This involves end-to-end deep learning approaches that directly map the camera input images to predicted maneuvers as well as classical approaches that construct a feature vector from object detection results (including several estimated properties of the detected objects) and apply recurrent neural networks on those features.

The works described above are based on Supervised Learning. In addition, we also investigate RL as an effective way to approach complex situations where acquiring a sufficient amount of high quality, diverse and correctly labeled training data is impractical or too demanding. We use several simulation frameworks where a vehicle learns to drive and navigate in different situations and environments. Our current investigations proceed along several lines:

- Drive and navigate collision-free on inner city streets with static obstacles
- Adjust the acceleration, deceleration and lane change behavior in the presence of several other cars on the same and neighboring lanes
- Analysis of knowledge transfer possibilities: use models that have been trained in relatively simple environments also in more complex environments and benefit from their already learned knowledge.

The focus of our studies is on DRL, especially modern Actor-Critic-methods where neural networks help in clustering complex environmental states into groups of similar states. This makes the mathematical computations on finding the optimal action for a given state possible for real-world AD applications.

The Future: Quo Vadis?

In this final section, we outline our expectations regarding which directions the future AI research and the respective development for AD will take.

The performance of ML-based perception approaches continues to improve

We expect further incremental improvements of the detection and classification accuracy of ML-based perception approaches just by further increasing the number and diversity of the used labeled training data. Despite the huge amounts of data already collected for many problems, the possible performance gains that can be realized by using more training data are not yet saturated. An example for this is the camera-based detection of pedestrians, where large databases are available since years, and new and larger databases have been made available recently (e.g., [37]), leading to better pedestrian detection performance. Still, an even better performance can be expected by using even more data [38].

In addition, the Neural Network architectures and ML techniques themselves are expected to improve further, mostly based on intelligently automated large-scale empirical analyses. Approaches like "Progressive Neural Architecture Search" [39] and related techniques that optimize the model structure will become more important and more commonly used. The general scheme is that besides the training of the model, also the model selection and the tuning of its meta-parameters (and those of the learning process) are automated. This requires significant computation resources and intelligent model selection and tuning processes, which in turn can be guided by dedicated ML techniques [40]. Thus, sophisticated ML approaches will be used to produce other, sophisticated ML models.

Effective Reinforcement Learning will become more important

In general, high quality, diverse and reliable data for training, validation and test of ML systems is required in tremendous amounts, which is a big challenge for the future. So far, there is no fully adequate replacement for largely manual or at least human-supervised generation of labeled data for many scenarios. One possible solution is RL, which can reduce the complexity and quality requirements for manually created training data. In recent years, RL has been very successful for complex, but "closed", well-defined problem domains. For instance, the software "AlphaZero" from "DeepMind" [41] reached very high playing skills beyond human abilities for the board games Go, chess and shogi, outperforming all previously existing software for those games. For "open" problems such as AD – where there is no clear, straightforward and unambiguous definition of success and failure – RL is so far not as successful, typically inferior to Supervised Learning in cases where enough labeled training data is present. However, RL for AD is currently receiving a lot of research attention (e.g., "Berkeley Deep Drive") and has a very high potential. We expect some breakthroughs in the medium-term future in this area.

Specifically DRL is of advantage in situations that we can simulate reasonably well. With modern simulation tools for automotive applications, this is the case even for relatively complex urban traffic scenarios. As was stated before it is quite intricate to conceive algorithms that handle the part of planning and decision-making in a robust manner in a large variety of real-world situations that are confronting us in AD.

Most current algorithms tackling this domain use many assumptions about the possible scenarios and need careful handcrafting with a lot of expert knowledge about written and unwritten traffic rules. This leads to complex, dedicated and hard-to-maintain software. Plenty situations can arise that are impossible to solve with the given ruleset as they were not anticipated during the development process. To attain a more generic approach that is applicable in a diverse set of traffic scenarios a DRL approach is possible. While in current systems, knowledge injection from external sources into the system through rules and guidelines is the norm, novel approaches using DRL gain knowledge and experience by facing a huge amount of simulated traffic scenarios. Using efficient training methods, these approaches can lead to deployable planning and decision-making modules that adapt to a broad range of different situations and scenarios without relying on expert knowledge. There is still a lot of room for further research in this domain and we expect some breakthroughs in the medium-term future in this area.

Another possible direction is the highly automated generation of new-labeled training data by AI/ML approaches. Once a very good AI system featuring largely trustable decisions is available, it can be utilized for such a task, extending its own operational domain gradually by moderate extrapolations to new data instances that are slightly out of its previous scope. However, the risk of "reinforcing" learned errors and thus strengthening error patterns or creating a systematic, adverse bias in its decisions is very high – in case the output and self-assessment (e.g., in form of classification confidences) of the AI systems should not be as trustable as assumed. This is a crucial barrier to overcome in future research.

Intelligent interaction of perception and planning

In typical current AD systems, the perception and planning components are largely independent. In some architectures, a high-level orchestration module determines which perception and planning activities are to be performed in a certain situation – based on manually designed rules. In addition, in this case, "sensing" and "planning" are executed in a purely sequential order, largely independent of each other (just appropriately configured by the orchestration module). We expect that this will change in the medium-term future: the interaction of the sensing and planning components itself will become "intelligent".

In particular, the planning module will feature a high degree of "situative awareness". It needs to "understand" the current environmental situation. Based on this understanding, its task is to not only draw the correct conclusions from the perception data and provide the correct input to the acting module, but also to control central aspects of these modules. Regarding the perception module, "attention" is such an aspect that will most likely be controlled. Despite the further increasing hardware resources that will be available in the future, we think it will be neither fully possible nor economical to create a complete, fine-grained and accurate representation of all aspects of the environment in each processing step. Instead, the perception module should focus its attention on those aspects and entities of the environment that are most important in the current situation and provide a detailed and reliable representation of them (while still devoting sufficient attention to other aspects of the environment, in order to be able to react to an unexpected change in the driving situation). Thus, conceptually, we expect a move of the design paradigm from independent perception and planning components to a "decision-aware perception" [42].

One line of promising research along this direction goes beyond pure object detection labels (e.g., where in the image is a pedestrian). It additionally incorporates labels of contextual information like relevant activities (e.g., pedestrian is standing, slowly walking, crossing the street, etc.) and interactions (e.g., pedestrian is or is not looking at the ego-vehicle, makes a head or hand gesture towards the vehicle, etc.) [43]. This can improve the prediction of the pedestrian's behavior, as ML approaches have the possibility to incorporate such contextual information early in their recognition process and thus link the appearance of the person with her or his predicted behavior in the next few seconds already in early perception stages. As the predicted behavior of the pedestrian is closely related to decisions, an autonomous car might need to make, such a "behavior-aware" pedestrian detection could become one aspect of a general "decision-aware perception" in the future.

The right balance between hand-written software, classical AI and ML

We expect that future intelligent AD systems will be neither large end-to-end systems, nor pure sensing-planning-acting systems that are enhanced with ML here and there. A key aspect for the further development of AI systems for AD will be the "right" balance between conventionally designed software (hand-written source code, including fixed rules, classical non-ML AI methods such as logical inference, search procedures, etc.) on one side and ML (classical methods, DL, DRL) on the other side.

In the near future, the highly automated vehicles that can drive largely autonomously in certain defined situations will most likely be based on the classical sensing-planning-acting paradigm. They will use ML (including DL) inside clearly distinguished and largely separated modules of the system – mostly inside visual perception modules like object detection, semantic segmentation, and road and infrastructure analysis. The planning and acting and the interactions between the sensing, planning and acting parts will be largely based on carefully designed rules (probably partially with parameters that were optimized using large-scale statistics and other empirical data).

In the medium-term future, step by step more modules will use ML (including DL and DLR) inside and the interactions between the modules and between the top-level parts sensing, planning and acting will be partially based on ML. The clear borders between perception, planning and acting will be relaxed, but they will not vanish. Aspects such as "saliency" (where to focus attention) and the awareness of the situation, the general context and the decisions that are to be made will become more important.

Even in the longer-term future, a full end-to-end DL or DLR system that is largely trained on (empirical or simulated) data as a whole (and that uses only a few hand-written rules and classical software algorithms) is not on the horizon. In addition, in the distant future, no complete end-to-end ML black box will control autonomous vehicles.

Thus, finding and optimizing this "right" balance will stay an important research topic also in the end. A big challenge is the interplay of manually designed strategies and behaviors learned by ML techniques. For example, basic traffic rules are probably best incorporated manually. The relative importance of different objects and their expected behavior in a certain situation can be learned to a large degree. ("Is this car on the left lane likely to cut in?"). In many cases, a combination of both is required. ("A ball just rolled onto the street. Thus, expect a child to run after it. Slow down and focus the attention on the most likely places where the child might appear."). We think that mastering this interaction will become a key issue for the long-term success of AD.

Verification and approval of AI systems

The biggest challenge for future AD systems is the verification of their correct behavior in all situations that can occur as part of their operational design domain, and the subsequent approval of their usage by normal car users. When the clear separation of sensing, planning and acting is still in place and ML is applied inside small modules only – as we expect for the near future – the test and verification of the overall system is challenging. However, verification is probably still feasible, as classical software test principles can be applied for most parts: the individual modules can be tested and verified in isolation, broadly independently of each other. Nevertheless, comprehensive testing of the small ML parts is problematic, as very large databases of all kinds of possible situations need to be used to verify their correct performance.

This problem is amplified in case sophisticated interaction patterns between sensing and planning components are used based on ML again – as we expect for the medium- and longer-term future. Significant parts of the system do not depend on each other anymore, thus they cannot be tested largely independently, but their interactions need to be tested already in early test stages. In the worst case, the testing effort could grow exponentially with the number of interactions, whereas in the best case, it might grow approximately linearly. Thus, minimizing the required verification effort will become a very important system design target. This is confronted with the future widespread use of ML techniques and intelligent interaction patterns as discussed above, causing a principle goal conflict.

Mastering this balancing act between the superior performance of complex AI systems on one side and their practical verifiability on the other side will become a dominant problem of AI-based AD systems. We expect that a huge amount of research and development effort will be dedicated to this issue in the future. If unsuccessful, we might experience another kind of new "AI winter": Sophisticated AI systems that show impressive performances for various use cases would be available, but could not be verified and approved for everyday use. However, if successful, the long-standing dream of fully autonomous driving would finally become true.

References

[1] S. Russell and P. Norvig, Artificial Intelligence - A Modern Approach, 2016.

[2] R. C. Arkin, Behavior-Based Robotics, 1998.

[3] R. Bellman, An Introduction to Artificial Intelligence: Can Computers Think?, 1978.

[4] P. H. Winston, Artificial Intelligence, 1992.

[5] E. Rich and K. Knight, Artificial Intelligence, 1991.

[6] D. I. Poole, R. G. Goebel and A. K. Mackworth, Computational Intelligence: A Logical Approach, 1998.

[7] R. S. Sutton and A. G. Barto, Reinforcement Learning: An Introduction, 1998.

[8] S. D. Pendleton, H. Andersen, X. Du, X. Shen, M. Meghjani, Y. H. Eng, D. Rus and M. H. Ang, "Perception, Planning, Control, and Coordination for Autonomous Vehicles," 2017.

[9] A. Swief and M. El-Habrouk, "A survey of Automotive Driving Assistance Systems technologies," 2018.

[10] A. Mousavian, D. Anguelov, J. Flynn and J. Kosecka, "3D Bounding Box Estimation Using Deep Learning and Geometry," 2017.

[11] X. Chen, K. Kundu, Z. Zhang, H. Ma, S. Fidler and R. Urtasun, "Monocular 3D Object Detection for Autonomous Driving," 2016.

[12] Y. Xiang, W. Choi, Y. Lin and S. Savarese, "Subcategory-aware Convolutional Neural Networks for Object Proposals and Detection," 2017.

[13] D. Feng, L. Rosenbaum and K. Dietmayer, "Towards Safe Autonomous Driving: Capture Uncertainty in the Deep Neural Network For Lidar 3D Vehicle Detection," 2018.

[14] B. Yang, W. Luo and R. Urtasun, "PIXOR: Real-time 3D Object Detection from Point Clouds," 2019.

[15] C. R. Qi, H. Su, K. Mo and L. J. Guibas, "PointNet: Deep Learning on Point Sets for 3D Classification and Segmentation," 2017.

[16] H. Gao, B. Cheng, J. Wang, K. Li, J. Zhao and D. Li, "Object Classification using CNN-based Fusion of Vision and LIDAS in Autonomous Vehicle Environment," 2018.

[17] D. Maturana and S. Scherer, "VoxNet: A 3D Convolutional Neural Network for real-time object recognition," 2015.

[18] C. R. L. W. Qi, C. Wu, H. Su and L. J. Guibas, "Frustum PointNets for 3D Object Detection from RGB-D Data," 2018.

[19] Z. Wang, W. Zhan and M. Tomizuka, "Fusing Bird View LIDAR Point Cloud and Front View Camera Image for Deep Object Detection," 2018.

[20] E. Arnold, O. Y. Al-Jarrah, M. Dianati, S. Fallah, D. Oxtoby and A. Mouzakitis, "A Survey on 3D Object Detection Methods for Autonomous Driving Applications," 2019.

[21] I. A. Tarmizi and A. A. Aziz, "Vehicle Detection Using Convolutional Neural Network for Autonomous Vehicles," 2018.

[22] G. Melotti, "Multimodal CNN Pedestrian Classification: a Study on Combining LIDAR and Camera Data," 2018.

[23] J. Schlosser, C. K. Chow and Z. Kira, "Fusing LIDAR and images for pedestrian detection using convolutional neural networks," 2016.

[24] D. A. Ridel, E. Rehder, M. Lauer, C. Stiller and D. Wolf, "A Literature Review on the Prediction of Pedestrian Behavior in Urban Scenarios," 2018.

[25] S. Jugade, A. Victorino, V. Berge-Cherfaoui and S. Kanarachos, "Sensor based Prediction of Human Driving Decisions using Feed forward Neural Networks for Intelligent Vehicles," 2018.

[26] J. &. C. H. Kim, M. Hwangbo, J. Choi, J. Canny and Y. Kwon, "Deep Traffic Light Detection for Self-driving Cars from a Large-scale Dataset," 2018.

[27] X. Li, J. Li, X. Hu and J. Yang, "Line-CNN: End-to-End Traffic Line Detection With Line Proposal Unit," 2019.

[28] W. Schwarting, J. Alonso-Mora and D. Rus, "Planning and Decision-Making for Autonomous Vehicles," 2018.

[29] T. Okuyama, T. Gonsalves and J. Upadhay, "Autonomous Driving System based on Deep Q Learning," 2018.

[30] T. P. Lillicrap, J. J. P. A. Hunt, N. Heess, T. Erez, Y. Tassa, D. Silver and D. Wierstra, "Continuous control with deep reinforcement learning," 2016.

[31] Y. Zhang, P. Sun, Y. Yin, L. Lin and X. Wang, "Human-like Autonomous Vehicle Speed Control by Deep Reinforcement Learning with Double Q-Learning," 2018.

[32] M. Kaushik, V. Prasad, M. Krishna and B. Ravindran, "Overtaking Maneuvers in Simulated Highway Driving using Deep Reinforcement Learning," 2018.

[33] P. Wang, C.-Y. Chan and A. de La Fortelle, "A Reinforcement Learning Based Approach for Automated Lane Change Maneuvers," 2018.

[34] W. Yuan, M. Yang, H. Li, C. Wang and B. Wang, "End-to-end learning for high-precision lane keeping via multi-state model," 2018.

[35] Y.-K. &. H. C.-Y. Lai, Y.-H. Huang, C.-W. Huang, Y.-X. Kuo and Y.-C. Chung, "Intelligent Vehicle Collision-Avoidance System with Deep Learning," 2018.

[36] M. Morsali, J. Aslund and E. Frisk, "Trajectory Planning for Autonomous Vehicles in Time Varying Environments Using Support Vector Machines," 2018.

[37] M. Braun, S. Krebs, F. Flohr and D. M. Gavrila, "The EuroCity Persons Dataset: A Novel Benchmark for Object Detection," 2018.

[38] D. M. Gavrila, "EuroCity Persons: A Novel Benchmark for Vulnerable Road User Detection (invited talk)," in *ECCV*, 2018.

[39] C. Liu, B. Zoph, M. Neumann, J. Shlens, W. Hua, L.-J. Li, L. Fei-Fei, Y. A. J. Huang and K. Murphy, "Progressive Neural Architecture Search," 2018.

[40] F. Hutter, L. Kotthoff and J. Vanschoren, "Automatic Machine Learning: Methods, Systems, Challenges," 2019.

[41] D. Silver, T. Hubert, J. Schrittwieser, I. Antonoglou, M. Lai, A. Guez, M. Lanctot, L. Sifre, D. Kumaran, T. Graepel, T. Lillicrap, K. Simonyan and D. Hassabis, "Mastering Chess and Shogi by Self-Play with a General Reinforcement Learning Algorithm," 2017.

[42] A. de la Fortelle, "The perception-decision gap (invited talk)," in *ECCV*, 2018.

[43] A. Rasouli, I. Kotseruba and J. K. Tsotsos, "Are They Going to Cross? A Benchmark Dataset and Baseline for Pedestrian Crosswalk Behavior," 2017.

Training and validation of neural networks in virtual environments

Raphael Pfeffer

T. Bertram (Hrsg.), *Automatisiertes Fahren 2019*, Proceedings,
https://doi.org/10.1007/978-3-658-27990-5_12

Trends in the Automotive Industry

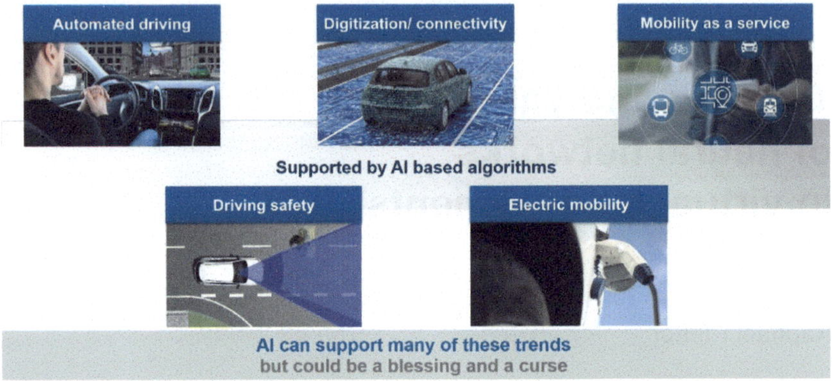

In recent years, artificial intelligence has increasingly become the focus of public attention. The automotive industry can profit greatly from this – especially in the rapid development of autonomous driving functions and driver assistance systems. In addition to established OEMs, newcomers to the industry are also investing large sums of money and are putting enormous effort into advancing this technology. Most big trends in the automotive industry can be supported by AI based algorithms.

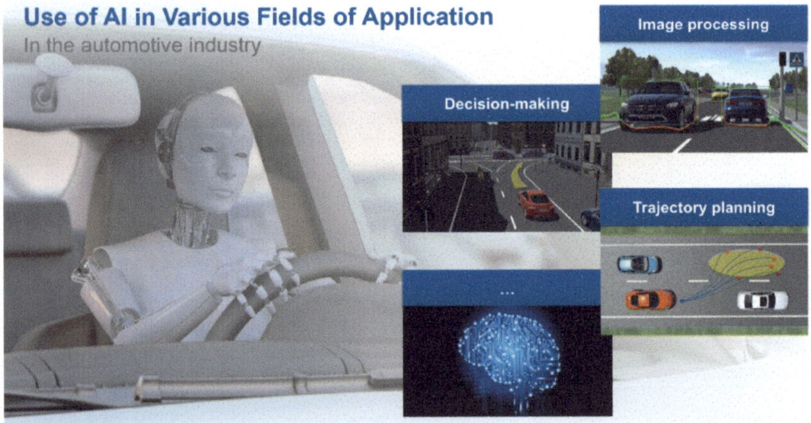

Functions Based in Neural Networks

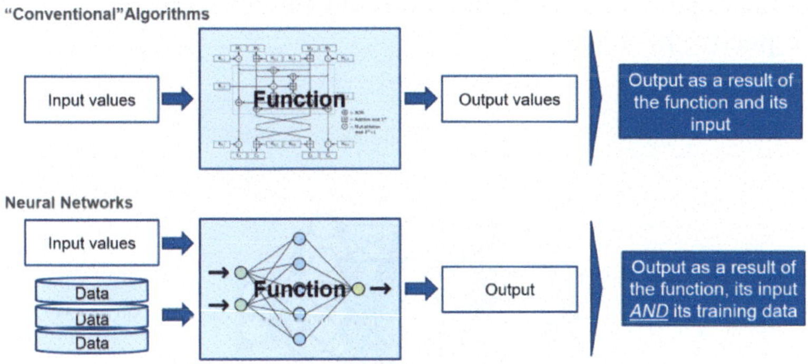

Two Main Phases of AI

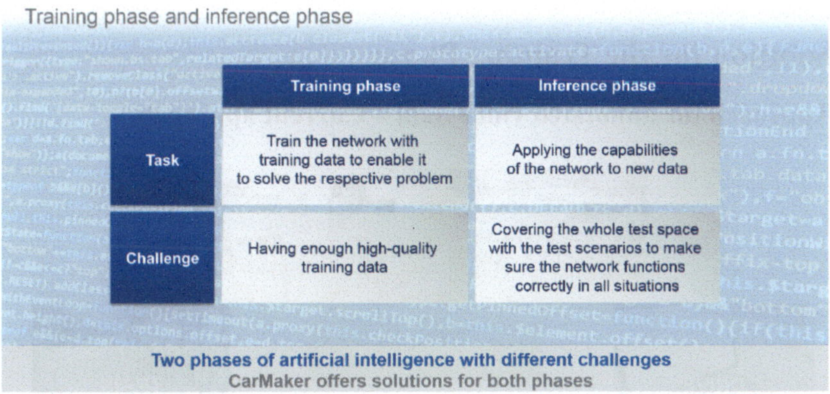

Training Phase – Algorithm/System Development

Train the network with training data to enable it to solve the respective problem

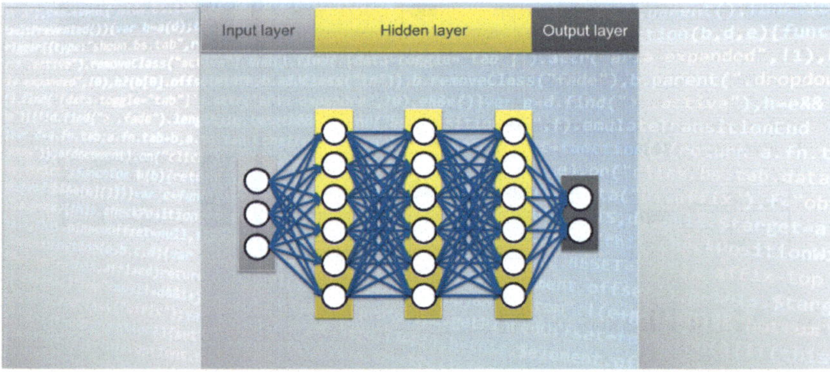

Solution for the Training Phase

Approach for annotated video data – Current approach

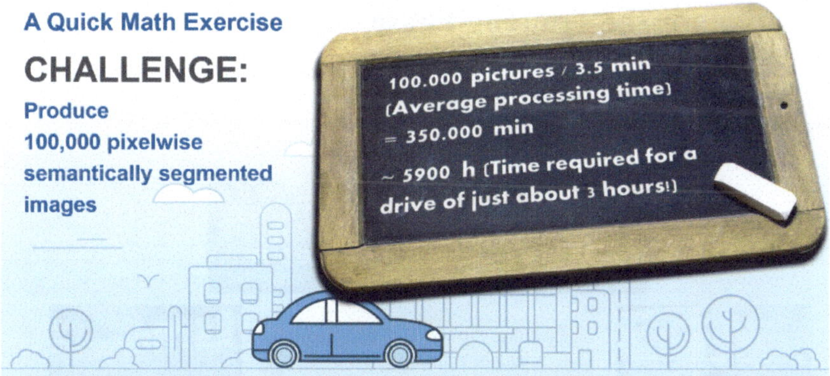

The quality of neural networks depends on the quality of the training data used. The neural networks only achieve high performance if the training data has been annotated thoroughly and as error-free as possible. In this example (picture above) it is the case that a traffic sign has not been annotated as such.

The quality of training data can be significantly improved if synthetic image data is used where the annotation of the data can be automated instead of being performed manually by humans.

Solution for the Training Phase

Approach for annotated video data – Potential for a better approach

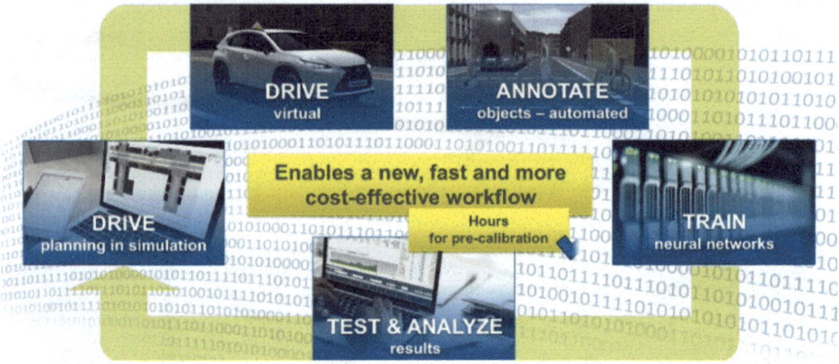

Open Integration and Test Platforms for Virtual Test Driving

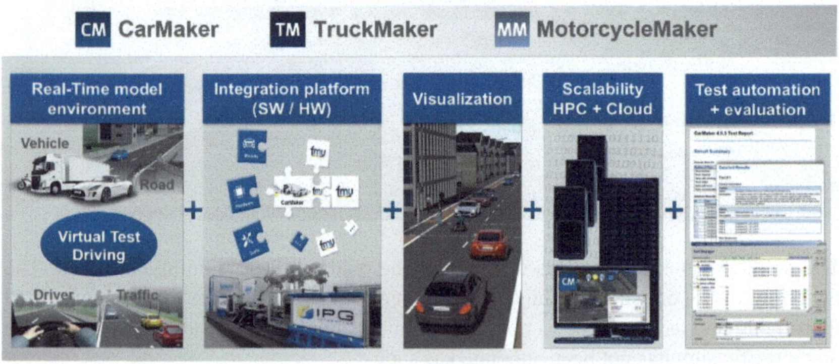

With the open integration and test platform CarMaker from IPG Automotive, the workload for annotation can be reduced. It can generate artificial (camera) data from a virtual environment, which can be created automatically together with the corresponding meta information.

Thus, a training data set can be prepared in the simulation, which is suitable for a training of the neural networks. Fine tuning can then be performed with a reduced data set of real images, which saves cost and time. In addition, CarMaker offers the possibility to integrate the trained deep learning algorithms in a closed loop environment and to test them simulatively in real driving scenarios.

Solution for the Training Phase

Generation of synthetic training data

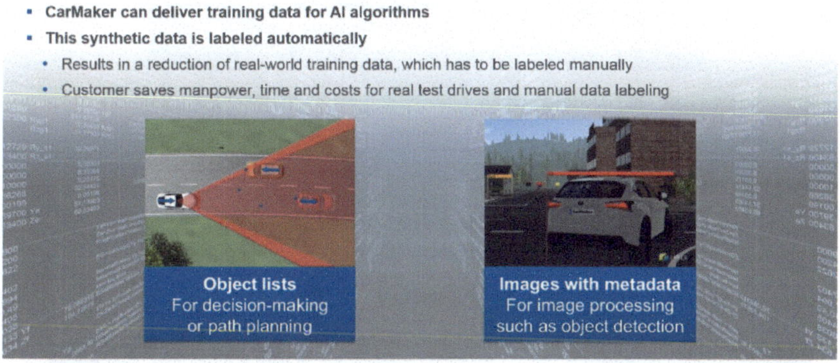

Example: Semantic Segmentation

Generation of synthetic training data

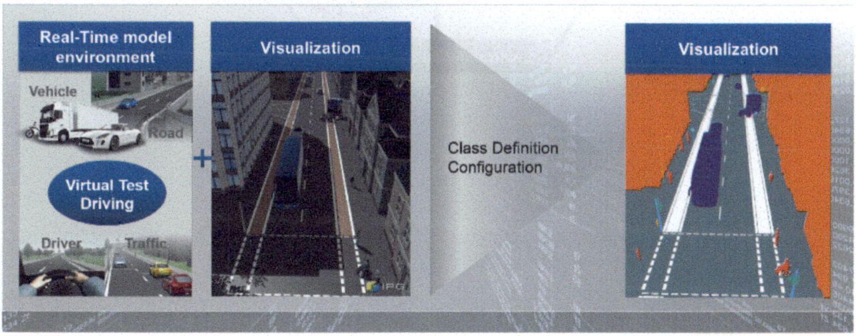

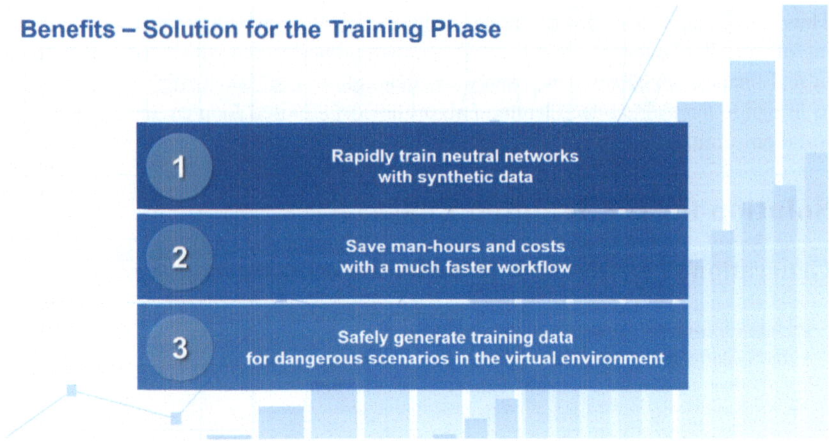

Testing in Real Scenarios – Solution for the Inference Phase

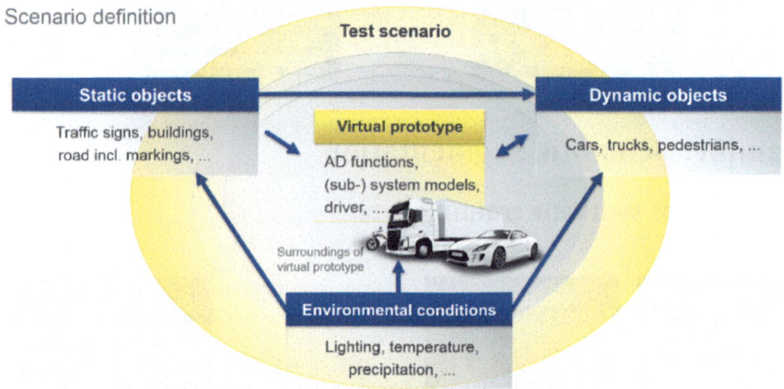

Integrating AI Algorithms into the CarMaker Environment

Many different interfaces available

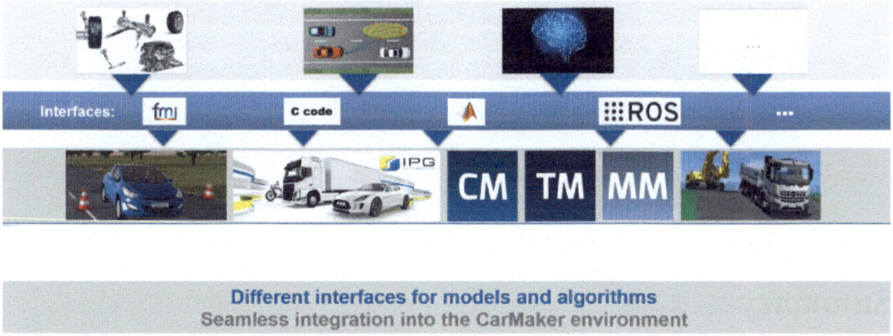

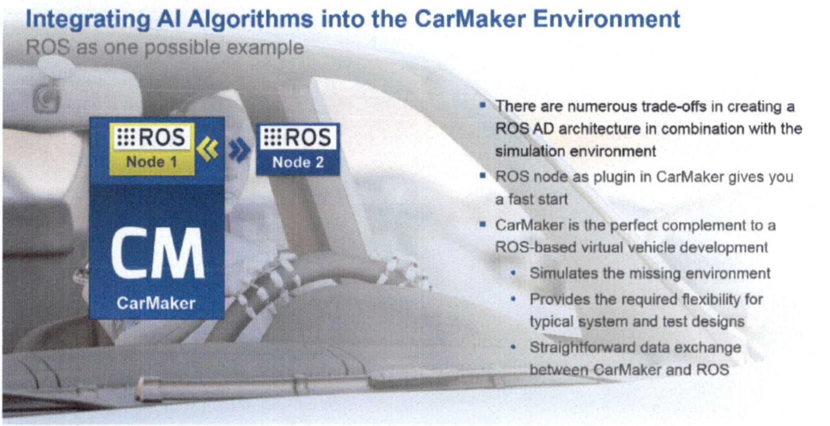

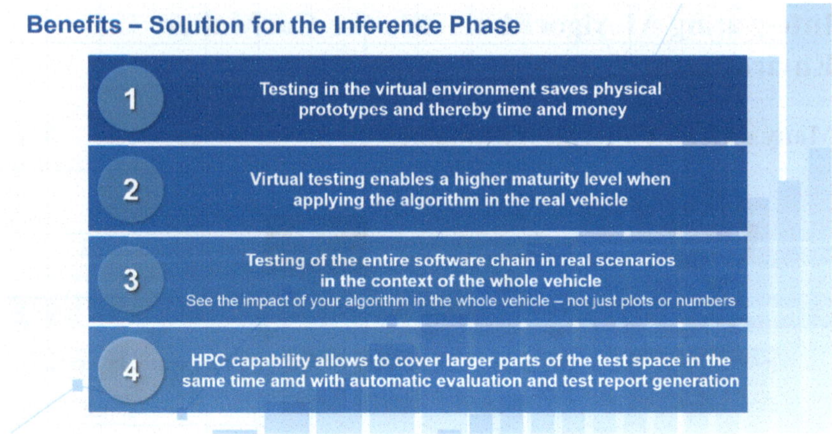
Benefits – Solution for the Inference Phase
1
Testing in the virtual environment saves physical prototypes and thereby time and money
2
Virtual testing enables a higher maturity level when applying the algorithm in the real vehicle
3
Testing of the entire software chain in real scenarios in the context of the whole vehicle
See the impact of your algorithm in the whole vehicle – not just plots or numbers
4
HPC capability allows to cover larger parts of the test space in the same time amd with automatic evaluation and test report generation

Summary

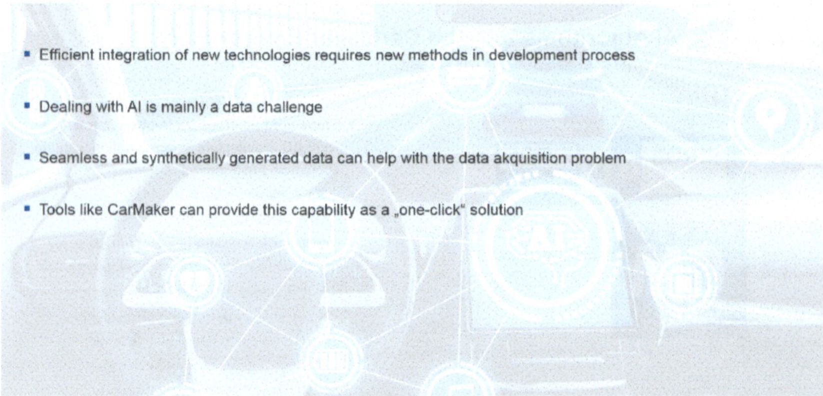
▪ Efficient integration of new technologies requires new methods in development process
▪ Dealing with AI is mainly a data challenge
▪ Seamless and synthetically generated data can help with the data akquisition problem
▪ Tools like CarMaker can provide this capability as a „one-click" solution

Methodology for the generation and execution of scenarios for the virtual driving test with automated driving functions

Martin Herrmann

T. Bertram (Hrsg.), *Automatisiertes Fahren 2019*, Proceedings,
https://doi.org/10.1007/978-3-658-27990-5_13

Introduction

It is now generally accepted that it is – to a reasonable extent – impossible to test and validate automated driving functions exclusively with the help of real driving tests. It certainly makes sense to feed measurement data from the field test into an SIL environment in order to test new software versions. However, the usable amount of data is by far not sufficient for a statistical proof of safety [1]. In addition, the data are no longer valid for other sensor or vehicle generations, or for a different system behavior caused by a new software version: The sensor data and the traffic behavior depend on the behavior of the system to be tested, but the control loop is not closed with this method.

It is therefore necessary to increase the test room coverage by further methods using the virtualization of the driving test and to investigate the behavior of the system or individual subsystems in a closed-loop control loop if possible.

In reality, there are numerous parameters that cannot be handled or can only be handled with difficulty. At this point, simulation offers the possibility to deal with at least some of these aspects:

- Large variety of traffic situations
- Country-specific features regarding road, environment and traffic
- Hardly reproducible traffic situations and environmental conditions
- Rare critical situations

This paper proposes a number of approaches aimed at systematically generating simulation scenarios for testing the automated driving function as completely as possible. Furthermore, it will be examined how the necessary test throughput can be generated and how the gap between simulation and application in the real world can be closed.

Scenario Concept

A scenario is a temporal representation of traffic situations and consists of the following elements: It is based on the street layout and the topology, including the course of the road, the number of lanes, road markings, etc. Building on this is the infrastructure, which contains all stationary objects such as traffic signs, vegetation, buildings or temporary changes such as construction sites.

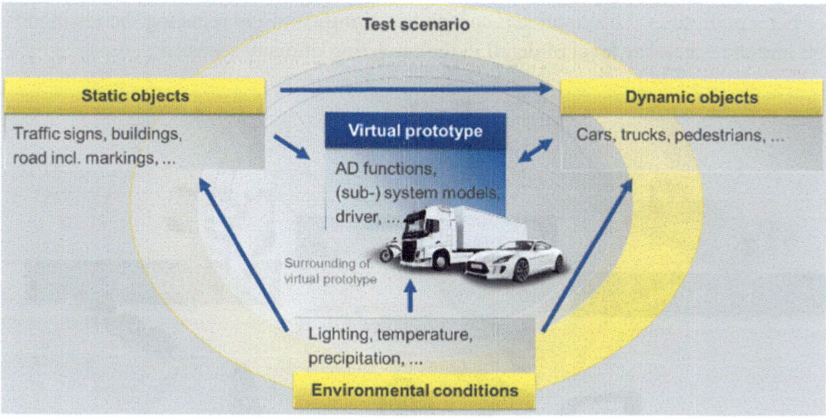

Fig. 1: Components of a test scenario [2]

Environmental conditions such as the position of the sun, precipitation or temperature are considered, too, which has a major influence on the function of the sensors in particular. The description of the motion of traffic objects such as cars, trucks, pedestrians, etc. completes the scenario challenging the automated vehicle in the simulation.

The Virtual Prototype

A validated vehicle dynamics model is the basis for the model of the automated vehicle, as it influences the sensor data by the movement of the vehicle body on the one hand, but also converts the steering and acceleration requirements into movements on the other. Critical driving situations close to the physical limits place correspondingly high demands on the modeling quality.

Also essential is the possibility of high-precision modelling of the sensors. For their realistic functionality, a detailed environment model is a basic requirement. However, individual elements of the automated driving functions can also be tested with simpler models, e.g. to rule out errors in perception or to investigate general behavior using statistical and phenomenological models.

“Functional decomposition" refers to the process of dividing a complete system to be tested into its functional components. Since the individual systems do not react equally sensitively to changes in the test parameters, they are initially secured independently of one another. [3] At this point, the virtual prototype in the open integration and test platform CarMaker from IPG Automotive offers the possibility to generate input data for the subsystems, for example object lists for maneuver planning or realistic image data for perception tests. This approach contributes significantly to reducing the testing effort and the necessary level of detail in the modeling of many scenarios.

Fig. 2 The Virtual Prototype

Scenario Generation

If the technical boundary conditions of the simulation are fulfilled, the main task is to develop an efficient methodology for generating relevant test scenarios on the basis of scenario databases. A combination of different sources (see Fig. 2) makes it possible to reuse existing data and to efficiently convert it into basic scenarios for simulation. A classification of the base scenarios ensures that specific groups of scenarios can be filtered out. Furthermore, system errors that are identified by means of suitable evaluation criteria can be quickly traced back to decisive boundary conditions.

In addition, suitable variation parameters and parameter spaces must be defined. This poses the greatest challenge, since knowledge of plausible values based on data from field trials provides a basis for the majority of parameters, but usually contains only comparatively few critical scenarios that lead to accidents in reality. One possibility is therefore a simulative investigation of extensive parameter spaces followed by a plausibility check of the movement of the individual road users. A physical modelling supports that no physically impossible scenarios find their way into the test database. In the following, different options are presented and their advantages and disadvantages are discussed.

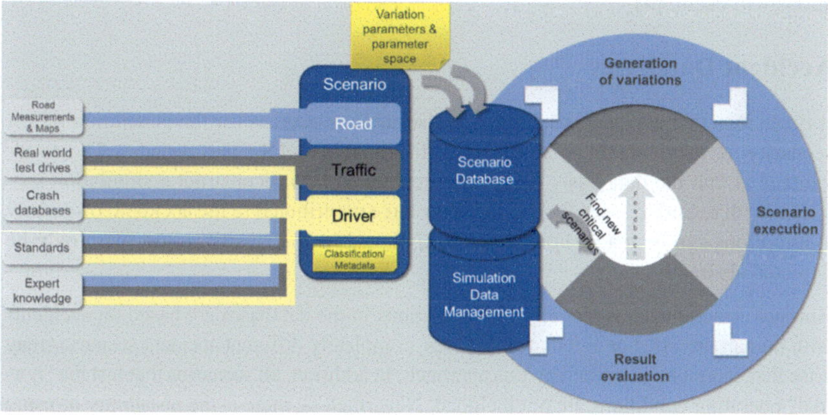

Fig. 3: Sources of scenarios and simulation flow

Generic event- and maneuver-based scenarios

A large number of possible traffic situations can be covered by a systematic generation of generic scenarios. A large number of variations can be generated automatically on the basis of experience values and the consideration of which typical basic scenarios can occur in the work area of the automated driving functions.

The disadvantage of this approach is that man as a limiting factor contributes to the fact that only traffic situations are included in this test database which have been considered and classified as relevant. For an automated driving function, other situations may be challenging than for the human driver. As a result, an apparently complete test database may not contain all relevant scenarios. [4]

Generic scenarios with driver behavior models

In contrast to event- and maneuver-based scenarios, scenarios based on driver behavior models do not require much effort to parameterize traffic objects. Different traffic situations result automatically from the interaction of the objects with each other and the traffic rules in the scenario. Theoretically, these scenarios can be executed for any length of time and new situations will always arise.

However, only interactions and misbehavior foreseen in the models are simulated – the unpredictability and randomness of the behavior of human road users is not or only to a limited extent taken into account. Consequently, only a few new, unpredictable situations will occur. [4]

Accident Databases

It is obvious that accident databases are also used as sources for the generation of test scenarios, as they contain very critical traffic situations of the past decades. The disadvantage is that the data basis is not complete enough to reconstruct a simulation scenario. With regard to some factors that may have contributed to the accident, often only assumptions can be made. For this reason, critical traffic situations can often only be depicted abstractly.

Another disadvantage is that all traffic situations in the databases are based on accidents with human drivers. For automated vehicles, completely different accident scenarios may arise that cannot be identified by this approach. In addition, all scenarios that test the "normal" operating conditions are not included. Nevertheless, there is the possibility of using challenging situations due to the misconduct of another road user for the validation.

Import of scenarios from the field test

One way to generate scenarios that are relevant for the automated driving function is to reuse recorded scenarios from the field test. The challenges here are the completeness and accuracy of the measurement data and the quality of the reconstruction in the simulation. In addition, there is the danger that humans will also have an influence on the recorded data, since they determine which traffic situations are considered critical or relevant and thus filter the data. In this way, potentially relevant scenarios are lost. [4]

The basis for the reconstruction of the scenarios is a list of the detected objects including their trajectories relative to the measuring vehicle over time as well as the trajectory of the measuring vehicle. This makes it possible to create a replay of the real traffic situation, as Ibeo Automotive Systems and IPG Automotive demonstrated back in 2016. [5]

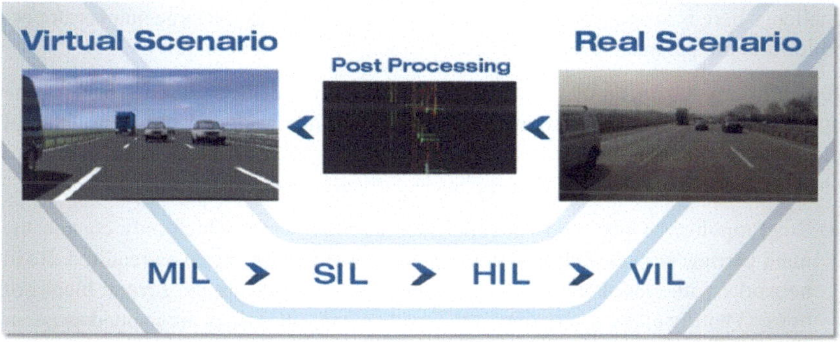

Fig. 4: Import of processed sensor measurements for re-use of real-world scenarios in the development process

Roads can be measured by service providers and built in simulation format or imported from HD map material. For simple environments (like motorway scenarios) it is also possible to reconstruct the road directly from the measured data, as long as the measured data contains information about GPS position, road markings, traffic signs in sufficient quality.

In further steps it is also possible to vary the traffic objects and the environment in the simulation in order to generate further similar scenarios and to further investigate the robustness of the system to be tested. Some of the options include:

- Variation of the road course incl. elevation profile, radius of curvature, etc.
- Removal or addition of traffic objects
- Change of object types and vehicle models
- Variation of relative positions lengthwise and crosswise

This procedure makes it possible to generate many very realistic scenarios with a data basis from the real world. The bigger the manipulations are, the more the question arises whether the scenarios are still realistic and therefore relevant for the system.

Execution of scenarios in virtual driving tests

Parameter variation and test automation

Regardless of the source of the scenarios, CarMaker offers the possibility not only to execute the basc scenario, but also to perform parameter variations to a greater or lesser extent, thus ensuring the robustness of the system by extending the test space. The challenge here is to apply suitable procedures in order to keep the large number of possible variation parameters and parameter characteristics under control.

Basically, any parameter of a basic simulation scenario, i.e. the static environment, the dynamic objects and the environmental conditions, can be varied. A clever structure of the scenarios ensures that plausible scenarios result even in the case of significant changes in parameter values. This applies in particular to the following points:

- Road topology: Open road networks such as motorways, which only contain the main carriageway as well as entrances and exits, can be varied as required. Road networks with closed streets (e.g. the mapping of streets around several blocks of houses) hardly allow a variation of the road course, since there are many dependencies between the individual streets.
- Traffic objects: If all actions and events relative to the ego vehicle are built up in maneuver-based scenarios, this enables the reproducible execution of scenarios independent of the behavior of the ego vehicle. Furthermore, it is ensured that a similar behavior to that of the basic scenario is achieved. If behavior models are used, the model parameters can be varied, but then it is no longer possible to predict the resultant traffic situations in advance.

Irrespective of the type of scenarios used, a full factorial investigation of the parameter space is also not practicable in the simulation. Here, evolutionary algorithms, gradient methods and other stochastic and target-oriented procedures can support the identification of system boundaries. Nevertheless, it should be ensured that a very large number of scenarios can be investigated in a reasonable time to support an agile development process.

Scalability

Depending on the existing hardware and purpose, the execution of the simulation can be scaled in different ways. On a computer with several CPUs or GPUs, for example, parallel calculations can be performed, while several of these computers can be used simultaneously within a network. For the highest possible performance, CarMaker can calculate a very large number of different processes in parallel using high-performance computing clusters, both locally and in the cloud.

On the simulation platform side, there are different approaches to parallelizing test catalogs. The most efficient variant is parallelization at the virtual driving test level. It offers almost perfect scaling and is a suitable approach for large test catalogs and many test kilometers.

Integration stages of the simulation platform

The test concept in the PEGASUS research project consists of simulating a very large number of scenarios in model-in-the-loop and software-in-the-loop environments and investigating identified critical scenarios with a higher level of detail in hardware-in-the-loop simulations or on the proving ground. [6]

The vehicle-in-the-loop method not considered in PEGASUS can help to minimize the gap between HIL and proving ground test. The vehicle is then completely integrated, only the sensor data is generated by the simulation. The real vehicle is on an empty test area and feeds its position back into the simulation. This makes it possible to carry out much more complex and dangerous scenarios than would be practicable in reality. In addition, many of the scenarios from earlier integration stages can be reused and reproducibly depicted, eliminating the need for a complex structure of the scenarios on the test site. [7]

Summary

For a proof of safety for automated driving functions, it is necessary to combine the methods mentioned in the paper and other methods not mentioned here in order to cover the test area as completely as possible. Research projects such as PEGASUS and Enable S3 provide approaches for possible security concepts taking into account some of these approaches.

The demonstration that these actually result in a safe product has not yet taken place. It is to be expected that further approaches to validation and scenario generation will have to be considered. The current methods enable the development of a scalable test and integration platform, the generation of a multitude of generic scenarios and the reconstruction of real traffic situations. Thus, fundamental functionalities for automated driving functions can be developed.

Nevertheless, further, unexamined situations will occur in the field test. It is therefore necessary to collect data in the real world by fleet tests for further system improvements. In this way, previously unknown scenarios arising from automation risks can also be included in the development and test process.

References

[1] W. Wachenfeld and H. Winner, "The Release of Autonomous Vehicles," in Autonomous Driving: Technical, Legal and Social Aspects, M. Maurer, J. C. Gerdes, B. Lenz, and H. Winner, Eds.: Springer, 2016, pp. 425–449.

[2] A. Höfer, "Purpose Driven Scenario Generation", Autonomous Vehicle Test & Development Symposium, Stuttgart, 2018

[3] C. Amersbach, H. Winner, „Funktionale Dekomposition – Ein Beitrag zur Überwindung der Parameterraumexplosion bei der Validation von höher automatisiertem Fahren, 12. Workshop Fahrerassistenzsysteme und automatisiertes Fahren, Walting, 2018

[4] P. Koopman, A. Kane, J. Black, „Credible Autonomy Safety Argumentation", 27th Safety-Critical Systems Symposium, Bristol, 2019

[5] R. Katz, "Automated Scenario Generation for Testing ADAS Based on Post-processed Laserscanner Data", Apply & Innovate, Karlsruhe, 2016

[6] T. Form, "PEGASUS Method for Assessment of Highly Automated Driving Function", SIP-adus Workshop 2018, Tokio, 2018

[7] M. Überbacher, P. Wolze, T. Burtsche: „ Experiencing Safety Function Testing", ATZ worldwide, Issue 7-8/2017

Solving the validation challenge of automated driving with a holistic test center

Simon Tiedemann, Andreas Mank

T. Bertram (Hrsg.), *Automatisiertes Fahren 2019*, Proceedings,
https://doi.org/10.1007/978-3-658-27990-5_14

The validation challenge: how much is enough?

Driver assistance and automated driving functions are already changing driving as we know it, and not simply by removing the need for human interaction. They are also making driving more enjoyable – and vastly safer. Thanks to significant improvements in software, hardware, and especially Artificial Intelligence (AI), cars will soon be able to think and make life-saving decisions completely by themselves. However, as we accelerate towards automated and autonomous vehicles, we are also seeing the emergence of new and unique challenges. In the end, the goal of assistance systems and self-driving software is to reduce the number of accidents and fatalities.

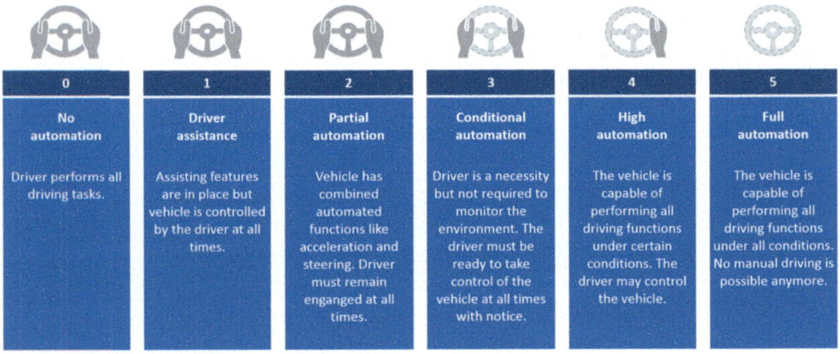

Figure 1:SAE Levels of automation

How much testing and validation is required to prove that automated vehicles are safe and ready for mass production?

This is the question that currently drives car manufacturers and suppliers all over the world. To gain acceptance by end users and authorities – we must first demonstrate, without question, that this new technology is reliable and safe. In the past, OEMs and Tier 1s have spent almost countless hours on the roads, testing advanced driver assistance systems (ADAS). Validating automated driving systems (SAE Levels 3 to 5) represents a far greater challenge – due to increased complexity, shrinking development cycles, and a host of new legal and safety requirements (Figure 1). [1]

SAE Levels 3 to 5 of automated driving require little or no human interaction. For this, automated vehicles must prove to be reliable in virtually every complex driving scenario. To achieve this, systems must be validated in a range far broader than ever before. Industry experts believe that to train and validate Level 4 and Level 5 functions, up to a billion of kilometers of data need to be collected to statistically experience the most diverse conditions and edge cases. [2] Several car makers are well along the way having collected billions of test kilometer data via their fleet of production cars. For actual testing of automated driving functions, the specific data required would take a lot of years to even collect in the first place.

In addition, further questions arise when talking about validating automated driving functions. Should we restart the entire validation process whenever parameter changes influence overall system safety? Where do we need to drive – and how many hours and kilometers are sufficient? How do we best distribute test driving across a range of different markets and evaluate all necessary testing situations? Automated driving functions contain unprecedented levels of complexity. As a result, more user-friendly solutions are needed to verify when the desired results are achieved. Persisting with current methods (such as those used to validate more basic driver assistance systems) is already proving impractical, since already today, test vehicles produce up to ten terabyte of data per day [3], making it a challenge to acquire and store this data efficiently.

The testing phase would be far too long, and the number of real-world test drive kilometers is simply unattainable. Given the strict requirements regarding the safety and performance of automated vehicles, the amount of discussion within the industry is understandable. Many companies individually but also consortia like the Pegasus project have been established to answer questions like: How many kilometers and hours of testing are required to properly validate these systems? At this point, no firm numbers have been agreed upon. Like mentioned before, leading OEM talk about billions of kilometers – it's obvious that on-road testing alone is not enough to achieve sufficient validation. To usher in the new era of automated vehicles, simulated test driving can, and must, represent most of automated driving evaluation going forward. At the same time, development and update cycles must decrease if automated vehicles are to reach the market within desired timeframes. Virtualization enables both the simulation and parallel replaying of test drive scenarios – accelerating and improving the entire validation process.

The validation toolchain

The answer to efficiently generate and make suitable driving data for test and validation purposes available lies in a comprehensive toolchain for automated driving that migrates a huge part of the required test kilometers into the virtual world. Through a suitable simulation method, the number of cost-intensive test vehicles and test drives can be significantly reduced. All of this is possible while also meeting legal safety requirements and significantly reducing development expenses.

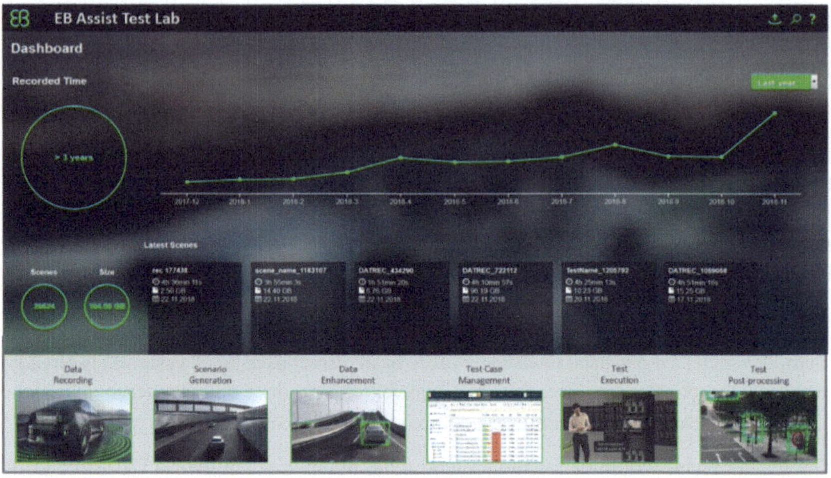

Figure 2: Validation toolchain for automated driving

In Figure 2, our approach to provide a comprehensive toolchain for validation of automated driving functions and systems is depicted. This toolchain will encompass recording and generating test drive data as well as a holistic platform to administer the acquired data and execute tests to allow for successful validation.

By leveraging this approach with a suitable cloud-computing platform, it becomes possible to easily scale to the power of thousands of high-performance computers to significantly reduce development time. The cloud offers a digital testbed that features global scalability, unparalleled flexibility to enable mobility companies to quickly test and analyze data, and thus go to market faster, eliminating the burden of infrastructure implementation and maintenance.

As an example, our partner Microsoft is one of the companies with a cloud network fast enough to meet the unique demands of automated driving. With data centers in more than 50 regions around the world, the Azure platform provides exactly what we need for this validation toolchain: a robust, intelligent, high-speed cloud network.

Within EB's automated test drive platform, there are several key functions which aid during different aspects of the validation process – these are described in the following.

Test drive recording

Car makers, Tier 1s, and technology providers are already amassing millions of test drive kilometers to build up their data libraries. By capturing and logging test drive data, algorithms can be tested, AI systems can be trained, vehicles and components can be validated, and higher fidelity simulation environments can be created. EB provides a suite of tools allowing our development partners to plan, organize, and monitor test drive recordings in a more efficient way. For test drives to be truly productive, a vast scope of driving and sensor data must be captured easily and reliably. Understanding this challenge, EB provides advanced data-logging software, automotive-grade hardware (which supports all common bus types and bandwidths) as well as a suite of tools required for direct annotation, triggering, and recording of feedback. With EB Assist CAR Box (Figure 3), we offer a reliable, high-performance automotive PC system for efficient data-logging, replay, and test drive simulation.

Figure 3: EB Assist CAR Box

Via this in-vehicle device, developers can easily capture all relevant vehicle sensor data – and store it ready for further processing. EB Assist CAR Box offers a robust in-vehicle solution used to capture sensor data under all types of road and weather conditions. With easy integrability into existing test vehicles, it provides a flexible input/output configuration, highly precise time-stamping with a resolution of 25 ns and accurately recorded sensor data to help accelerate and improve the development process. But how is all this data then transferred into the development environment? The answer lies in data ingest solutions like the Microsoft Azure Data Box, disk solutions that enables the reliable and inexpensive transfer of terabytes of data to the cloud.

Built within the overall toolchain, EB Assist ADTF (Automotive Data and Time-Triggered Framework), EB's tool for automated driving development, combines a development environment with an interactive work environment. Developers can use the graphical user interface and existing modules to create new configurations without writing a single line of code. They can use drag and drop to define the data flow between software components and execute tests immediately to see the effects. Faster development of automated driving software modules is facilitated by the libraries and toolboxes for various functionalities, which can be easily integrated into the framework. These functional interfaces and data formats are open to developers, and the framework is compatible with Microsoft Windows and Linux operating systems.

EB Assist ADTF simplifies the process of transferring captured test drive data into the development environment and allows validation to be performed on each individual sensor. Data can be archived in the cloud along with local attributes such as weather, road, and driving conditions. AI algorithms can be trained against the processed data, and metrics can be made of the entire test drive fleet. Led by a customer forum of leading car makers and Tier 1s, EB Assist ADTF is one of the most widely used ADAS development and test environments in the world.

Scenario generation

To get the vast amount of data for test and validation, only recording the data using a test drive fleet is simply impossible. Thus, we are powering an industry shift towards virtual and simulated vehicle testing. Not only does this significantly increase the scope and range of test coverage – it reduces the costs and timelines of the entire validation process. In addition, virtual testing is already compatible with existing vehicle sensor sets. Using computer simulation, data from a range of different sources can be stored in a master database. This can include vehicle accident data, field operational tests, driving simulation studies, traffic simulations as well as general industry knowledge. This broad scope of data can be captured, logged, categorized, and then extracted into the most relevant testing environments.

To run successful test drive simulations, manufacturers first need the ability to generate accurate driving scenarios. With precise test drive information captured via EB Assist CAR Box, existing driving scenarios, as well as entirely new ones, can be simulated. Our partner ecosystem provides the simulation tools needed to generate realistic driving environments by populating individual assets into large HD maps or procedurally generated layouts. Capturing accurate test drive data ensures physically correct modeling of the car's behavior and the behavior of other objects – as well as the realistic simulation of vehicle sensors.

Data ingest and storage

Thanks to their various onboard sensors, cameras, and LIDAR systems, test vehicles have the capability to generate tens to hundreds of terabytes of data per day, for use in validation and development. However, these vast amounts of data must be ingested and stored in a suitable development environment before they can be processed. Compounding this big data problem is the fact that test vehicles are often nomadic. They do not return to the same garage every night – since the following day's testing may be completed in a different location – adding to the ingestion and storage challenge. As test fleet locations continue to change and grow in number, timely data ingestion can be extremely difficult as well as tracking and traceability poses a challenge. Building infrastructure at the edge is an inefficient and cost-prohibitive approach and managing networking logistics across multiple locations is also a challenge. These challenges are the reason why cloud-based solutions have become a critical component of data ingestion and storage for development purposes. Data storage appliances are designed to offer bulk imports of terabytes of data from any location to the cloud for faster processing.

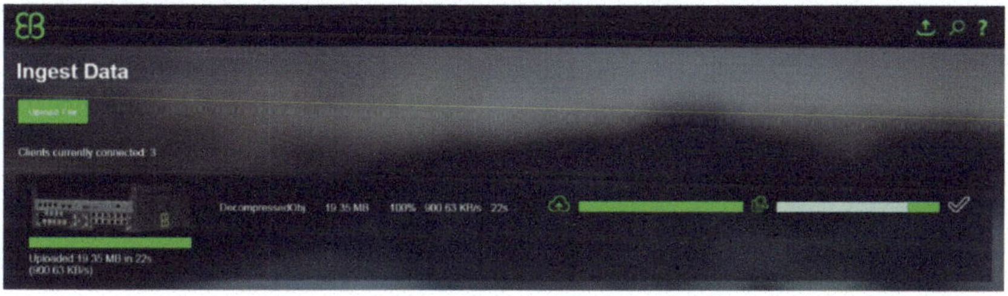

Figure 4: Data ingest from EB Assist CAR Box

Data enhancement

To manage increasing amounts of test drive data in a cost-effective manner, EB's software includes high levels of data automation. The ability to transmit data to the cloud is supported, for example, with disc logistics and upload facilities. Basic meta data can be generated automatically as new vehicle recordings are uploaded. AI-based services used in the generation of meta-data (such as driving context) are also supported. Additionally, EB offers a full-blown SDK which allows customers to easily add new analysis modules to extend the existing functionality.

EB's (AI-based) tools also enable the efficient, semi-automatic labeling of test drive recordings. This significantly increases labeling speed and reduces costs as an alternative for manual labelling service companies. With petabytes of ingested data, real-time data processing and analysis is critical for this stage. Various analytics services will be available to fully manage scale-out clusters for achieving the desired scale and speed of analysis.

Test case management

The described toolchain also enables the generation and administration of test cases and test sets. Users can easily define preconditions such as the software version and algorithm state. Input data can be listed and managed – such as recordings and simulation scenarios (including variations). Software can provide expected results such as ground truths and KPIs. The toolchain also provides the automatic generation of test cases – as well as the detailed history and coverage of previous test cases. EB Assist ADTF supports multiple use cases and numerous applications, from predevelopment to series projects and furthermore provides easy data exchange. On top of that, the many available add-ons and interfaces for various use cases (e.g. electronic horizon) and other well-established tools and standards (e.g. MATLAB Simulink, AUTOSAR) all allow for accelerated and flexible development and the execution of tests.

Test execution and post-processing

EB enables the execution of abstracted test cases to cover any kind of test scenario, both hardware-in-the-loop (HIL) and software-in-the-loop (SIL) in a transparent and efficient way. Via the incorporation of a suitable cloud platform, EB Assist ADTF supports nearly infinite scalability and the execution of parallel tests. Several functions allow developers to conveniently and efficiently perform their tasks:

- Improved efficiency of existing hardware by providing smart load balancing
- Decreased down time by automatically detecting and handling execution errors
- Efficient post-processing of tests and test runs
- Automatic creation of meaningful test reports
- Administration and archiving of test results
- Comparison between test results

As an example, within this toolchain, each data package is automatically scanned and enriched with meta data – such as weather and road condition information. Via this method, developers save the time and money associated with searching for suitable driving scenes – as the data is already correctly labeled. In a practical example (Figure 5), we validated a grid-fusion software component. Via a grid-based simulation of the vehicle's surroundings, the system provided concise information about obstacles and obstacle-free areas. The test center then compared the output of the grid-fusion algorithm with a set of reference data. KPIs for occupied, non-occupied, and unknown areas were compared. The test-data was then loaded into the cloud, executed simultaneously, and processed further.

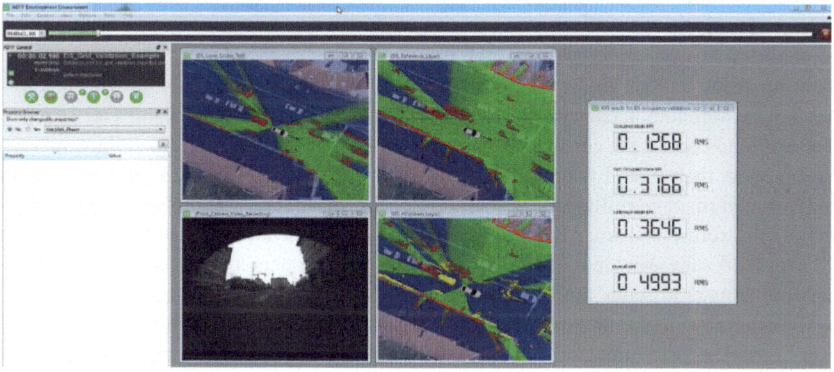

Figure 5: Example of grid validation

Overall, this unique cloud-based approach enables users to manage an increasing amount of test drive data whilst keeping costs low – thanks to automated labeling and data enhancement. Furthermore, it significantly accelerates the management, execution, and post-processing of parallel tests – validating automated driving functions quickly and efficiently. And lastly, leveraging the cloud allows for flexible and scalable testing, making it easier to meet critical peak demands.

An open design to suit every developer

The interfaces within the described automated driving validation toolchain are specifically designed in an open and accessible format. Via this approach, our development partners can add or customize their own services – using the entire toolchain, or only selected parts. This includes automatically generating meta information, enabling ground truth labeling, and performing tests on powerful 'in-the-loop' systems for specific validation tasks. Additional solutions can also be built by combining EB's and other existing tools with additional partners – such as simulation experts, test tracks, or data plugin specialists. Furthermore, there is scope to extend beyond the currently presented approach. For instance, test drives can be further improved by creating meta data based on speech recognition, or overall test kilometers needed for development, test, and validation purposes can be reduced using a smart process to select data. Overall, we are continually working to further close the gap between recorded data and simulated data.

Conclusion

Getting ahead of data challenges now is key to avoiding potential issues within automated and autonomous driving development in the future. Car makers and Tier 1s must be careful to choose the correct path forward when deciding how best to collect, ingest, and process their data. EB provides a suite of proven-in-use hardware and software to capture and manage millions of kilometers of test drive data – allowing developers to efficiently refine and improve their automated driving systems. Cloud solutions like Microsoft Azure provide a powerful platform to developers around the world to efficiently manage millions (and soon billions) of kilometers of test drive information as well as petabytes of simulation data. Already collaborating with many of the world's top car makers, our toolchain will be making driving more reliable – and more enjoyable – for all road users.

Bibliography

[1] National Highway Traffic Safety Administration (2018): Automated Vehicles for Safety. Society of Automotive Engineers (SAE) Automation Levels, https://www.nhtsa.gov/technology-innovation/automated-vehicles-safety, accessed 18.12.2018.

[2] Ohnsman, Alan (2016): Toyota's Robot-Car Line In The Sand: 8.8 Billion Test Miles To Ensure Safety, https://www.forbes.com/sites/alanohnsman/2016/10/03/toyotas-robot-car-line-in-the-sand-8-8-billion-test-miles-to-ensure-safety, accessed 18.12.2018.

[3] Accenture (2018): Autonomous Vehicles: The Race Is On, available at: https://www.accenture.com/us-en/insights/communications-media/autonomous-vehicles-data-challenges.

Toolbox for test planning and test realization of scenario-based field tests for automated and connected driving

Thomas Otto, Rico Auerswald
Fraunhofer Institute for Transportation
and Infrastructure Systems (IVI)

T. Bertram (Hrsg.), *Automatisiertes Fahren 2019*, Proceedings,
https://doi.org/10.1007/978-3-658-27990-5_15

Abstract

Current test and validation procedures for automated and connected driving functions use field tests on public roads primarily to identify unknown critical scenarios. These scenarios are then validated or varied within simulations or on closed proving grounds. In order to validate complex urban driving scenarios, this approach may need to be supplemented by the method of scenario-based field tests in real traffic environments. Scenario-based tests constitutes the methodical basis for an efficient and purposive test of automated and connected driving functions in public testbeds.

This paper focuses on concepts for the design, planning and implementation of scenario-based tests of automated and connected driving functions under real traffic conditions. The developed toolbox will be explained using the example of the “Digital Testbed Dresden / Saxony" as part of the initiative “Synchrone Mobilität 2023”.

Introduction

Synchrone Mobilität 2023

The "Synchrone Mobilität 2023" initiative aims at advancing intelligent transport systems (ITS). Focusing on automated and connected driving in urban areas, the research and development projects under its umbrella initiate scientific and technological developments and provide comprehensive test facilities. Beyond that, they offer manifold networking and cooperation opportunities for regional, federal and international partners. Financially supported by the automotive industry in Saxony, the projects receive additional funding from the European Union, the State of Saxony and the German Federal Ministry of Transport and Digital Infrastructure (BMVI).

The focus of the activities are:

- Digital testbeds in Saxony with a variety of road infrastructures as open test environment
- Automated and connected driving functions in the context of urban traffic flows and smart traffic control
- Self-driving public transport and commercial vehicles
- Vulnerable road users – connecting bicycles and pedestrians to the traffic system
- Reduction of CO_2 and pollutant emissions
- Increased traffic efficiency and safety

There is a cooperation among more than 50 partners from industry, research and public authorities with heterogeneous competences and responsibilities.

All partners are organized in multiple research projects with complementary subjects.

Testbed Dresden

The success of automated and connected driving in urban areas requires not only simulations and test drives on closed test sites but also enormous testing effort under real traffic conditions. The Digital Testbed Dresden offers optimal conditions for a variety of different traffic scenarios. The initiative "Synchrone Mobilität 2023" analyzes the road infrastructure in Dresden on the basis of requirements for testing automated and connected driving functions as well as forthcoming innovative mobility concepts and supports the development of an appropriate connected test infrastructure.

In the Dresden Testbed, there exists an established network on C-ITS topics of connected and automated driving between Fraunhofer IVI and the local stakeholders, including the City of Dresden and the State Office for Roads and Traffic of the Free State of Sachsen. In close cooperation with the latter two authorities, Fraunhofer IVI plays currently the key role in the specification, procurement, installation, operation and maintenance of the Dresden Testbed's network of roadside units (RSU). The Dresden Testbed – outlined in Fig. 1 – is located in urban and suburban spaces of Dresden.

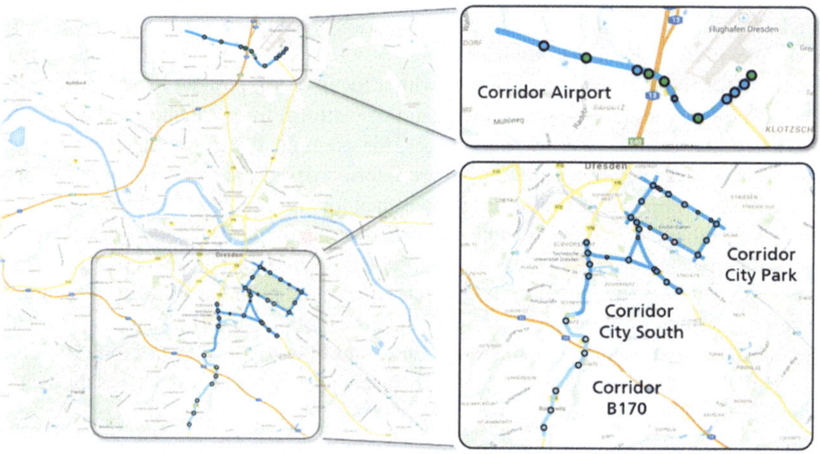

Figure 1: Overview of Dresden V2X-Testbed in 2019

Corridor Airport: urban and suburban road, approx. 6 km, 50 to 70 km/h, two lanes, divided roadways, separate cyclists' lanes, connected to highway, 5 RSU in operation, further 6 RSU planned

Corridor City South: urban road, approx. 8 km, 50 km/h, two lanes, partly divided roadways, separate cyclists' lanes, incl. more complex junctions, tram line, tram tracks and crossings, 3 RSU in operation, further 8 planned in 2019

Corridor City Park: urban road, approx. 8 km, 50 km/h, one or two lanes, partly divided roadways, complex junctions, bus lanes, tram tracks and crossings, approx. 12 RSU planned in 2019

Corridor B170: rural road, approx. 5 km, 50 km/h to 70 km/h, two lanes, divided roadways, separate cyclists' lanes, connected to highway, hilly profile, 7 RSU planned in 2019/20

Backend and Infrastructure for Automated & Connected Driving

Our C-ITS infrastructure outlined in Fig. 2 comprises three main blocks, namely the central backend services, the decentralized RSU, and the connected (test) vehicles. The central cloud hosts the backend services required for the operation of the testbed, including monitoring and the management of various resources. The latter aspect is developed within the project Resource Management System for Highly Automated Urban Traffic (REMAS), among other things, with a focus on support-services to utilize the Dresden Testbed as a test environment and test platform for automated and connected driving. Additionally, external data sources, for example, the traffic management system VAMOS [1] deployed in Dresden are interfaced and the information is pre-processed. Where applicable, this information is then forwarded to the decentralized clouds deployed at the respective RSU. Other backend services aggregate and process the information generated from every RSU and the test vehicles. The connectivity between the entities in the testbed and the central cloud is established via cellular communication, and, in the case of some RSU, via directional radio. The RSU, hosting the RSU clouds and directly connecting the traffic light controller to this clouds, employ a range of technologies for the communication with the connected vehicles as well as the central cloud services.

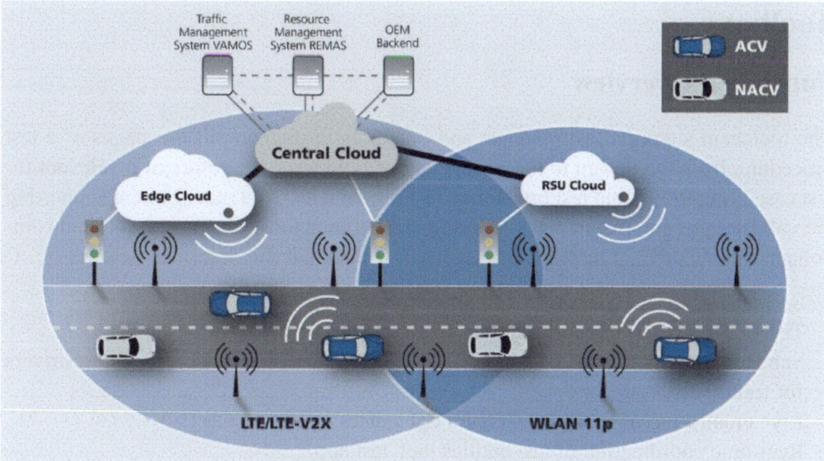

Figure 2: Backend and Communication for automated & connected vehicle

The central computing platform is necessary for services ingesting data from a larger geographical area, such as, centralized monitoring and management solutions, or applications interfacing external services. Both computing platforms should be scalable with respect to the available resources and allow a modular deployment of C-ITS services. Regarding the V2X communication, a wide range of existing and emerging technologies should be supported, such as, IEEE 802.11p, cellular data, or cellular V2X. As it is unclear whether a single or multiple communication technologies will prevail, the developed C-ITS services should not be tied to a specific V2X technology.

Due to the fact that the Dresden Testbed is comprised of public roads, it should be noted that over the near future it is very likely that other connected vehicles, which are not participating in any test activities, driving on the test corridors and generating V2X signal noise. This has to be considered in the system design in the sense that, depending on the communication channel used, during testing V2X messages may be received by and from non-participating vehicles.

Toolbox

Functional Overview

The toolchain supports test planners and operators throughout all the stages of a test procedure, from the design to the conducting of field tests. Therefore, it is relevant for test case generation, field test procedure, test runs and test data management supporting the validation of automated and connected driving functions. The most important components and features of the field test toolchain are named and described below:

- Scenario editor, validation and simulation framework
- Test driving path recommendations, exportable to multiple test vehicles
- Scheduling of available communication infrastructure, test vehicles and test drivers for test drives
- Live monitoring of test drives, traffic light states and RSU
- Real-time coordination of test entities incl. test app
- Database of recorded data from test vehicles and infrastructure components

The coordination and support of test runs with complex maneuvers driven by multiple automated and connected vehicles in urban areas is a priority task of the toolchain. Therefore, we developed a test-planning tool for the preparation of vehicle-specific test routes and situative driving instructions. The actual test runs are supported both by a monitoring function as part of the trip management and a test app which supports the safety driver in following the scenario sequence. With measurement data from test runs, it is also possible to reconstruct driving scenarios and vary them, e.g. for simulation purposes (traffic simulation, driving simulation).

Standardization

Standardization is in particular required to enable compatibility and interoperability of test scenarios. This is essential, considering the growing number of possible and indispensable test cases for automated and connected driving functions, which have to be interchangeable between different test environments as well as between industry, public authorities and research. Therefore, it was also a requirement for the interfaces of the toolbox to be compatible with existing standards. For the test of automated and connected driving functions, these are especially the ASAM standards OpenDRIVE and OpenSCENARIO.

OpenDRIVE

“ASAM OpenDRIVE defines a file format for the precise analytical description of road networks. Unlike other file formats typically used for navigation systems, ASAM Open-DRIVE's main use is in the area of simulation applications, which require exact road geometry descriptions, including surface properties, markings, signposting and logical properties such as lane types and directions.” [2]

OpenDRIVE forms the basis of HD-maps for testing automated driving functions. Due to the high density of details within the map and the accuracy of the track model, it will be possible to allocate driving scenarios to the real traffic environment precisely as well as to plan test routes including the calculation of driving trajectories.

OpenSCENARIO

“ASAM OpenSCENARIO defines a file format for the description of the dynamic content of driving and traffic simulators. The primary use-case of OpenSCENARIO is to describe complex, synchronized maneuvers that involve multiple entities like vehicles, pedestrians and other traffic participants. The description of a maneuver may be based on driver actions (e.g. performing a lane change) or on trajectories (e.g. derived from a recorded driving maneuver). Other content, such as the description of the ego vehicle, driver appearance, pedestrians, traffic and environment conditions, is included in the standard as well." [3]

The standard describes vehicle maneuvers within sequences, which are linked to specific start-, end- and cancel-conditions. Those trigger-events can be, e.g. the exceeding of a defined speed or the reaching of a defined relative distance to a vehicle ahead. The standard’s data scheme enables the exchange of driving scenarios for automated and connected driving functions between various test environments, like simulations, driving simulators, closed proving grounds and real traffic environments. Therefore, OpenSCENARIO is considered in this toolbox.

For the market introduction of automated driving functions, their hedge and release are currently key challenges. Scenario-based tests are one way to solve these problems [4]. The requirements towards a scenario description, transferring the test space into a scenario-depending representation to enable the comparison of scenarios across test domains are anything but simple. [4] tries to classify the terms in context of scenario-based tests. An approach for a description which gets close to the requirements is OpenSCENARIO [5].

Use Cases

Dresden Testbed focuses on the seamless integration of connected, automated and conventional vehicles in mixed traffic scenarios within city areas. In order to accomplish this goal, innovative basic services as well as advanced features are provided by a heterogeneous cloud and communication infrastructure. Some use cases are for example (outlined in Fig. 3):

- Cooperative lane change maneuvers (A)
- Turn maneuver alignment to avoid blocking situations (B)
- Predictive intersection crossing and traffic light approaching for blocked flow-offs (C)

To consider safety, driving comfort and energy consumption aspects for automated driving functions in urban traffic, functionalities for anticipatory and cooperative driving will be supplemented, taking other road users and traffic light controls into account (e.g. [6]). Dresden Testbed also aims at a novel approach on traffic optimization by providing maneuver-based driving recommendations for vehicles or vehicle swarms via connected urban road infrastructure.

The optimization of traffic flow is possible by combining intelligent vehicles with intelligent infrastructure facilitating anticipatory and cooperative driving. Further use cases outlined in Fig. 4, considering in particular driving lane recommendations and urban platooning.

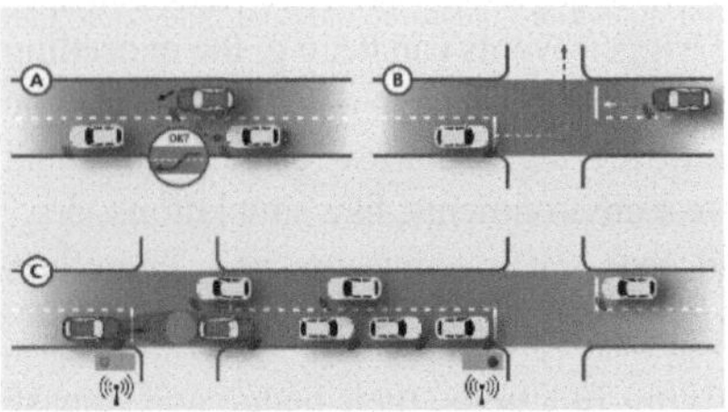

Figure 3: Use Cases of automated & connected Driving.

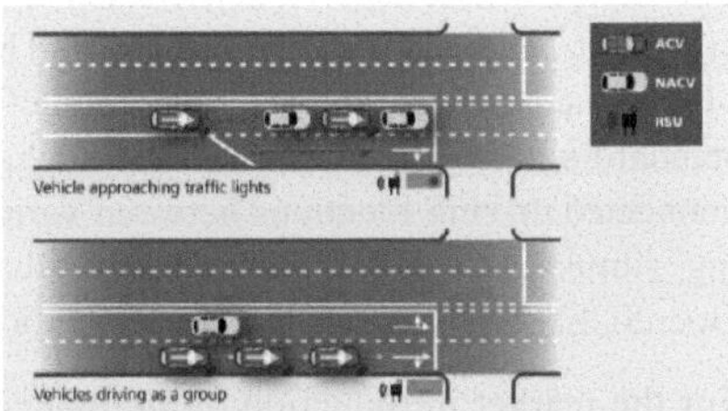

Figure 4: Further Use Cases of automated & connected Driving.

The first demonstration runs took place in mid-November 2018. It was a milestone in the Saxon overall initiative “Synchrone Mobilität 2023”.

Exemplary Procedure

The workflow and functionalities of the toolbox will be demonstrated with the use case "cooperative lane change maneuvers (A)".

Scenario Editor

The first step is to define a functional scenario description. The objective of a test case for the cooperative lane change is to validate that a cooperation between two vehicles with the help of V2V communication is achieved. Driving on a multilane road, one vehicle intends to change lane, while another vehicle on the neighboring lane is blocking the cut-in area of the vehicle with the lane change intention. Hence, the lane-blocking vehicle have to behave cooperatively and create a gap.

The test case includes additional constraints despite the driving sequence itself, e.g.:

- At least two (automated and connected) vehicles
- Divided (sub-)urban roadway
- More than one lane per direction
- Speed limit above 50 km/h, up to 70 km/h

The second step is to look for fitting track and traffic conditions for the chosen scenario. Therefore, there are two alternative approaches to choose from. Firstly, it is possible to use queries for comparable and already existing test scenarios to receive typical locations and conditions for the requested test scenario – if an adequate scenario database is available. Secondly, queries over the road segments of the real traffic environment are also feasible. For this exemplary test case we used the second approach, mainly because infrastructural requirements can be well defined.

Therefore, we use the OpenStreetMap database for a rough classification of the road features. This can be e.g. road category, number of lanes, speed limits, intersection types the equipment of intersection with traffic lights as well as the existence of a physical separation of oncoming tracks can be selected (outlined in Fig. 5).

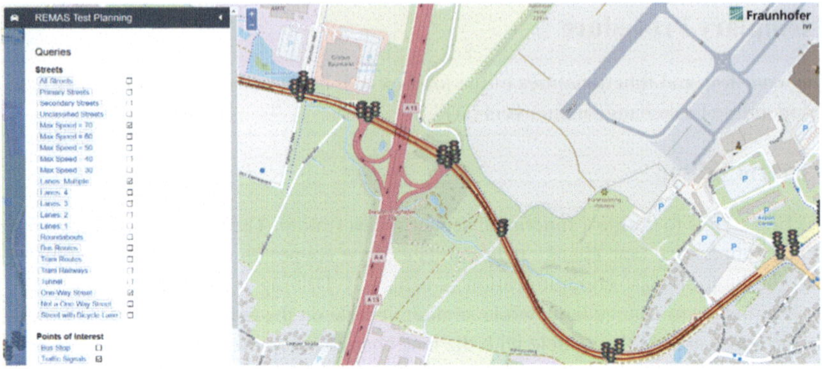

Figure 5: Snapshot of the test planning tool – in this case to identify a relevant road segment for the cooperative lane change test case.

More specific road features, like road widths, road markings, traffic signs, guardrails, guide posts or different road characteristics such as inclines, slopes, cambers or surface textures, can be supplemented by a parsing of the OpenDRIVE HD-maps on selected road segments (outlined in Fig. 6). The complexity of the traffic scenario often correlated with the traffic conditions. Therefore typical traffic conditions (e.g. traffic volume) of the different tracks and times are stored in the database.

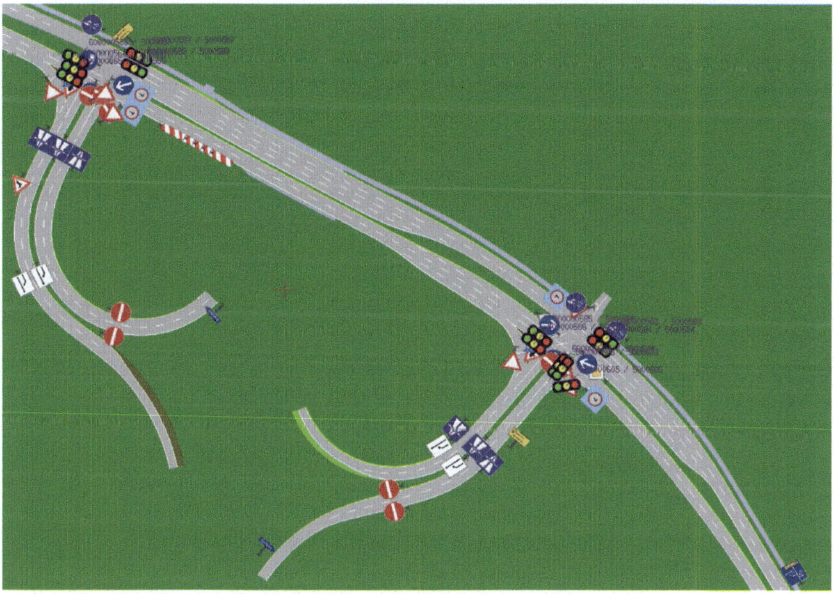

Figure 6: HD-map of a road segment within the Digital Testbed Dresden.

After relevant road segments have been identified, a suitable test route have to be set and situative driving instructions with their trigger-events have to be defined. For this purpose two alternative approaches can be adequate, depending on the available data basis:

1. Adapting comparable test scenarios, e.g. from simulation or a test scenario database.
2. Creating new test scenarios for the selected track.

The two approaches will be typically accomplished within a scenario editor. In both cases the scenario sequence have to be (re-)simulated under the perspective of a safe execution in public road traffic. Additionally, test scenarios in a real traffic environment will always vary more or less from the intended sequences. A parameter range have to be defined in which a conducted test run will be assigned to the test case. For the exemplary test case of a cooperative lane change a short excerpt of constraints could be, e.g.:

- Start and completion of the scenario within the defined geofenced road segment
- Acceptable speed range of 50 to 70 km/h (considering a 70 km/h speed limit on the test route)
- Acceptable acceleration of -3 to 3 m/s^2
- Acceptable time headway of 2 to 5 s
- Acceptable execution time for lane change maneuver of 5 to 15 s

A deviation will result in a cancelation of the test scenario. With a full set of constraints it is possible to generate a list of acceptable concrete scenario sequences. These are then transferred to an in-vehicle application of the vehicle under test (VuT) and (one or more) fellow vehicles.

Depending on the test requirements, additional processing can be added on top of the generated test sequences. One example is the approach of [7] for an OpenSCENARIO-based precise generation of trajectory. Scenarios in the form of OpenSCENARIO and road-networks in the form of OpenDRIVE are combined to derive global target trajectories for every dynamic scenario entity with a special regard to an application on proving grounds. Future work will also focus on an automated route planning task based on road geometry and lane attributes. This is non-trivial, especially when adapting it to a test bed in public traffic. Additionally, [8] show the challenges to transfer functional requirements from ISO 262626 to a scenario definition which should also be considered in the test preparation.

Scheduling

The list of acceptable concrete scenario sequences will be linked with the test case for the cooperative lane change. Requirements on resources for the test case can either be defined or derived directly from the linked scenarios. Relevant test resources are besides the VuT, e.g. communication infrastructure (RSU- and backend-services), connected fellow vehicles and test drivers. The scope of resource-scheduling depends on the test type which subdivides into naturalistic and controlled test runs [9]. For example, in order to conduct controlled test runs, relevant actions from the concrete test scenarios have to be assigned to the respective fellow vehicle of the test case.

In case of the cooperative lane change maneuver, another connected vehicle is required additionally to the VuT. Due to the lacking support of V2V connectivity in present real traffic, the controlled test approach should be picked which accompany the allocation of a connected fellow vehicle as the cooperation partner. In our example, the VuT should execute the lane change maneuver in some test runs whereas in some other it should create the gap for a cooperative lane change. Therefore, the VuT as well as the fellow vehicle need the driving instruction from both sides.

Real-time coordination of test entities including the test app

All test vehicles (VuT and fellown vehicles) are equipped with an in-vehicle test app. The respective test cases, containing the concrete test scenarios from the scenario editor, are transferred to the test apps. Because of the uncertainty when driving in public traffic, two modes for the test app are recommended:

1. Master-slave configuration
2. Equally balanced configuration

In the first mode, the VuT will be defined as master. When the actual driving maneuver differs from the projected one, e.g. because of surrounding traffic, slave-defined fellow vehicles will strive a compensation by adjusting their own maneuvers. With a balanced configuration on the other hand, also the situative driving instructions of the VuT will be adjusted in case of sequence-deviations.

While driving the actual test case, each test driver is given the individual situative driving instruction – both visually and acoustically. In the current version of the test app, the maneuver events are triggered time-based or location-based (outlined in Fig. 6 and Fig. 7). In addition, certain situational or relative events (e.g. distances between vehicles or communicated driving intentions) will be considered in the next release.

Figure 6: Coordinated Test Drive within the Digital Testbed Dresden.

Figure 7: Test App Visualization within the Digital Testbed Dresden.

As shown in Table 1, a multitude of event-trigger is necessary to fulfill the test sequence. Especially, event-trigger based on (V2X-)communicated information can be a key requirement for test cases covering cooperative maneuvers. Some of the relative triggers or even the actions themselves can be adjusted dynamically depending on the prior phases of the test sequence.

Table 1: Simplified excerpt of the test sequence for a cooperative lane change with the VuT in the role of performing the lane change and a fellow vehicle (FV) which is providing the gap using V2V communication.

Step	Test vehicle	Trigger	Actions*
1	VuT	Absolute: GNSS-Position of VuT	– Keep right lane – Set speed: ~65 km/h
	FV	Absolute: GNSS-Position of VuT (from V2X-communication)	– Take left lane – Decrease distance to VuT: ~3 m (blocking lane change)
2	VuT	Relative: Distance to FV (from V2X-communication)	– Communicate lane change intention – Activate indicator
3	FV	Relative: Lane change intention of VuT (from V2X-communication)	– Decelerate and increase distance to VuT: ~30 m – Confirm cooperative lane change
4	VuT	Relative: confirmation cooperative lane change from FV and distance to FV: > 20 m (from V2X-communication)	– start and proceed lane change within 15 s

**The shown parameter sets serves only as a sample in order to illustrate the trigger-action relation of the test case*

As a result, the test app controls the sequence of the test run, corrects the position, speed or the maneuver-timing of the test vehicles or aborts a test scenario if specific parameters can no longer be achieved. It also supports the test driver identifying whether the automated driving function operates like intended or not. This allows efficient testing in a real world testbed. The number of test drives is optimized, the time required can be minimized.

Analysis and Evaluation

Immediately after conducting a test scenario in a real traffic environment it should be verified that the actual driven test scenario matches with one of the intended maneuvers aggregated with the quantity of concrete test scenarios within the test case. If there is not a hit, the maneuver has to be repeated.

At the end of the test session the measurement data of the test vehicles as well as the data from the connected road infrastructure will be synchronized and merged within one data set. First short-analyses are then conducted, e.g. with regard to data consistency.

Conclusion

There are already a number of several tools focused on ITS-topics to develop and simulate test scenarios as well as components for test planning and test realization. Others are in the progress of being developed. However, these tools are widely used separately.

Due to the cooperation of different test entities as well as roadside- and backend-based infrastructure via different communications technologies, a powerful integration of these tools and components for an overall ITS-framework, plus the application as a real traffic environment is technically possible for the first time. This enables the basic requirements for a novel test approach that combines the advantages of realistic environmental conditions (such as the complexities of road traffic and road infrastructure) on the one hand and controlled test procedures on the other hand. The digital testbed cannot only be used for naturalistic field tests due to this approach. It represents also a suitable test environment for scenario-based testing. The challenges are complex. In particular, testers but also developers require tools for scenario-based testing in real world testbeds enabling them to receive controlled test procedures. The developed toolbox is intent for the purpose of "Digital Testbed Dresden / Saxony" and designed for complex scenarios, but it is also fundamentally transferable to other test beds. The above mentioned gearing of the "Digital Testbed Dresden / Saxony", the ITS test backend, test planning and test realization are basic components to address the entire sphere of activities.

Especially the manufacturer-independent planning and realization of tests requires enormous effort in the scopes of planning, coordination, realization and evaluation. In particular, the focal points of real-time test coordinations were previously restricted to single vehicles or simplified scenarios with limited interactions of one VuT mostly on closed proving grounds.

This resulted in tools primarily for a VuT-centered test context, which are extremely important for working on a proving ground and simulation of, for example, certain driving functions. However, these tools are only of limited use for testing especially cooperative driving functions under realistic conditions in an urban testbed.

References

[1] J. Krimmling: "Das Dresdner Verkehrsmanagementsystem VAMOS," in Verkehrsmanagementzentralen in Kommunen: Eine vergleichende Darstellung, M. Sandrock and G. Riegelhuth, Eds. Wiesbaden: Springer Fachmedien Wiesbaden, 2014, pp. 157–197.

[2] https://www.asam.net/standards/detail/opendrive/ (as consulted online on 11.02.2019)

[3] https://www.asam.net/standards/detail/openscenario/ (as consulted online on 11.02.2019)

[4] Steimle, Markus; Bagschik, Gerrit; Menzel, Till; Timo Wendler, Jan; Maurer, Markus: „Ein Beitrag zur Terminologie für den szenarienbasierten Testansatz automatisierter Fahrfunktionen." AAET 2018 – Automatisiertes und vernetztes Fahren, 14. bis 15. März 2018, Braunschweig

[5] Groh, Korbinian; Kuehbeck, Thomas; Fleischmann, Benjamin; Schiementz, Mark; Chibelushi, Claude: „Towards a Scenario-Based Assessment Method for Highly Automated Driving Functions". In: 8th Conference on Driver Assistance, 22 – 23 November 2017, Munich, Germany.

[6] Kloeppel, M., Grimm, J., Strobl, S., and Auerswald, R. (2019). Performance evaluation of GLOSA-algorithms under realistic traffic conditions using C2I-communication. In Nathanail, E. G. and Karakikes, I. D., editors, Data Analytics: Paving the Way to Sustainable Urban Mobility. CSUM 2018, volume 879 of Advances in Intelligent Systems and Computing, pages 44–52, Cham. Springer International Publishing.

[7] Heinz, Aaron; Wolfram Remlinger, Johann Schweiger: „Track- / Scenario-based Trajectory Generation for Testing Automated Driving Functions" 8. Tagung Fahrerassistenz, München, 22.-23.11.2017

[8] Till Menzel; Gerrit Bagschik; Markus Maurer: "Scenarios for Development, Test and Validation of Automated Vehicle" IEEE Intelligent Vehicles Symposium, 2018

[9] FOT-Net consortium, 2017, D5.4 Updated Version of the FESTA (Field opErational teSt supporT Action) Handbook, Version 6, http://wiki.fot-net.eu

Fusion of raw sensor data for testing applications in autonomous driving

Julius von Falkenhausen, Qi Liu

T. Bertram (Hrsg.), *Automatisiertes Fahren 2019*, Proceedings,
https://doi.org/10.1007/978-3-658-27990-5_16

To achieve the goal of an autonomous driving vehicle, more and more sensors are being integrated in the vehicle. For the last generations it was sufficient, that those sensors did send object data, which was then fused into an environment model. However, in order to have a more accurate model and to be able to navigate autonomously, the raw sensor data is needed. To test those sensors and the resulting environment model, a software is required, which can reliable fuse raw data into a high precision environment model.

There are three necessary key features, which are necessary so the framework can be used for different testing purposes. It needs a unique data model, which allows an easy integration of different sensor technologies (RADAR, LiDAR, Cameras) from different suppliers. It is also important to calculate uncertainties for every data point depending on all boundary conditions. These uncertainties have to be taken into account when creating the environment model and are used to determine the quality of the model at all stages. Since we are looking at recorded data, we can peek into the future and use the time independent information to analyse and improve the accuracy and predict the position and movement of objects, even at the time when the sensors originally couldn't detect them.

We propose a framework to handle the arising problems in analyzing the large amount of data in future ADAS architectures. The paper gives an overview of different modules and exemplarily shows one processing model.

1 Introduction

The testing and validation of autonomous driving functions gets more and more complicated with each step toward the fully autonomous vehicle. In the past it was enough to tailor a testing suite for each project, but with the increasing complexity of newer vehicle architecture this is neither efficient nor cost friendly. Therefore, we propose a modular approach to create a reference model for all testing scenarios.

First we will take a look at the different sensors and the raw data they deliver. Then we will focus on the different fusion models and give an example of a simplified signal processing model. We outline the challenges that arise in fusing raw data, which will lead us to the approach that we propose for testing and validation in the field of autonomous and assisted driving.

2 State-of-the-art

2.1 Sensor Technology

Radio Detection and Ranging (RADAR) has been widely used for obstacle avoidance in autonomous driving systems. By constantly scanning the objects using radio waves, Radar is capable of determining the distance information between sensor and objects. Compared to the camera and Lidar systems, which demand high dependency on the operating environment, Radar indicates high robustness under varying lighting, temperature and humidity conditions. The sensors with different detection ranges can be utilized depending on the system requirements, for example, long range Radar for Adaptive Cruise Control, medium range Radar for cross traffic alert and short range Radar for autonomous parking. However, it is difficult to directly employ Radar data for object recognition and classification. As well, the limited field of vision restricts its application in object detection in side direction.

Light Detection and Ranging (LiDAR) is another kind of important sensor in autonomous driving systems. Many of the Lidar sensors are applied to provide three-dimensional images with 360-degrees visibilities. The Lidar point clouds reveal accurate information and high resolution compared to that from Radar sensor. However, since the light spectrum waves could be seriously interfered due to unfavourable working environment, Lidar systems illustrate obvious shortcomings when operating in the scenarios of bad weather conditions, such as snow, rain, fog or the air full of dust particles. The excessive expense and large physical size are additional problematic issues in terms of wide application in vehicles.

Due to the improvement of image processing technologies in recent years, cameras have gained great attention in the field of advanced driver assistance systems. Conventional on-board cameras were implemented as the tool of monitoring blind spots. Nowadays, the camera-based object detection approaches have been used to recognize both moving and static objects including vehicles, pedestrians, buildings and traffic signs. Stereo cameras are desired in some scenarios requiring accurate depth information. Nevertheless, large effort has been necessary to calibrate the system. Monocular cameras, on the other hand, show advantages of low cost and compact size with easy calibration, while relative more complex methodologies are demanded to estimate distance information. Similar to the Lidar sensors, the main limitation of camera-based systems lies in the sensitiveness with respect to lighting and weather conditions.

In addition to the aforementioned sensors, the conventional motion sensors, such as ultrasonic sensor, optical encoder, gyroscope and GPS, also play important roles in motion estimation, path planning, simultaneous mapping and localization. Due to the high requirements on accuracy and reliability, one of the challenging tasks regarding the design of advanced driving assistance systems is to fuse sensor data from different types of sensors. Several attempts have been conducted on the advanced vehicle platforms. Companies such as Tesla were focusing on the development of vehicles with cameras and Radar sensor [1], while Google and Baidu prefer to fuse the result of Lidar, Radar and camera sensors [2] [3]. In order to have a comprehensive understanding of our raw sensor data fusion model, a brief introduction to the data fusion techniques will be given in the next section.

2.2 Related Work on Sensor Data Fusion

The topic of sensor data fusion attracts a lot of attention in the automotive applications. Many methodologies have been proposed to solve concrete tasks, which has been outlined in the review papers [4] and [5]. In recent years, one of the commonly accepted data fusion classification framework was defined in [6], which mainly covers three fusion models: raw data fusion, feature fusion and object fusion.

The raw data level fusion is applied to generate single resolution data by combining raw data from different sensor sources. The goal of this type of fusion methods is to improve spatial resolution, enrich synthetic data information, and derive highlighted data changes from multi-temporal sources. The structure of raw data fusion is illustrated in Figure 1(a). The raw optical data fusion approaches, which relies on substitution fusion techniques, modulation based fusion techniques, and multi-resolution analysis based fusion techniques, have been summarized in [7]. As well, the raw data fusion gains great attention in automotive applications. The researches have been carried out for vehicle detection based on the low-level data from the Radar system and a monocular camera [8]. Further approaches have been proposed using the low level measurement vector fusion from Lidar and Radar systems for a crash alert application [9]. The raw data fusion shows the advantage that the data can be classified at an early stage and the outputs of the fusion provide highly accurate data information [6]. Thus, invalid data can be eliminated, which leads to precise object detection and tracking results. Nevertheless, the raw data fusion requires extensive data bandwidth and intensive computational burden. These properties increase the difficulties of raw data fusion on the real-time system due to the limitations of computation capabilities, data volume and data format, especially for complex systems with multiple sensors involving Lidar, Radar and cameras.

Unlike raw data fusion that directly uses the sensor data as the inputs to the fusion model, the feature level fusion tries to extract certain features from raw sensor data and then maps the features for further data processing, which is shown in Figure 1(b). The common features that can be extracted include the edges, corners, lines and texture parameters. By extracting features from raw data, the data bandwidth can be reduced allowing for faster fusion approaches [5]. The applications of feature level fusion have been demonstrated in automotive applications based on Lidar and camera [10] [11] and Radar systems [12].

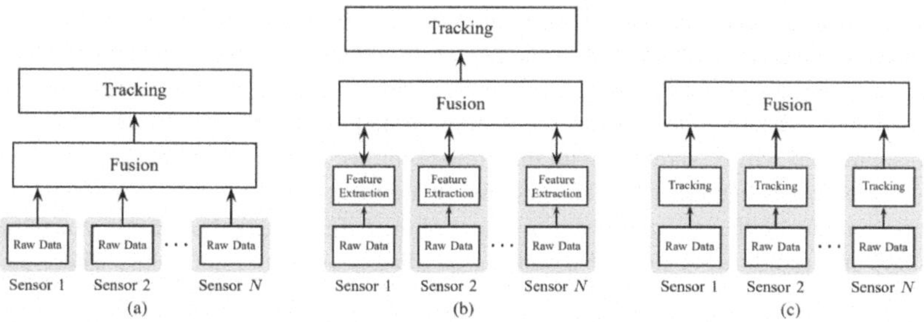

Figure 1. The structure of raw data level fusion system (a), feature level fusion system (b), and object level fusion system (c) [6].

Object level fusion is another fusion approach, which is to fuse lists of objects detected by different sensors. The architecture of the object level fusion is presented in Figure 1(c), where the detection and classification outputs of independent sensors are fused at the decision level for final classification. The multi-sensor based object fusion can prevent the problems of incomplete measurements of single or multiple sensors due to technical fault, system error or unexpected noise from severe operating conditions. Since the fusion procedure has been carried out at a high level, which relies on the object lists derived from tracking modules, this turns out to be a promising approach in real-time systems that require high simplicity and modularity. However, the sensors illustrate significant differences in terms of reliability. Thus, the problem arises that the object classification becomes more difficult due to the lack of associated raw data information regarding the relevant objects. Consequently, to optimize the object fusion result, it is necessary to design the system by utilizing sophistic tracking module [6] or by combining it with lower-level feature/raw data fusion approaches [13].

2.3 Challenges to raw data fusion in testing and validation

Current fusion models concentrate on the usage in the car and have different boundary conditions from the ones in validation and testing. While in the car there is a limited amount of computation power and time to process the data, in testing and validation we can use a lot more resources and time to achieve a higher accuracy. Furthermore, we can utilize the recorded object data across the timeframe to interpolate the movement of objects beyond the observed time frame..

One main problem for raw data fusion is the spatial and temporal association of the data from sensors of heterogonous technologies. A camera image, for example, gives information about the colour of a point with a known direction and unknown distance. The camera will deliver the information across its frame at the same time. LiDAR sensors have the missing distance information although at a different mounting position and therefore a different origin at which the angles are being measured. The LiDAR does not give information of all the points across the frame at the same point in time. Points are gathered over a certain timeframe. Assuming that at the end of the timeframe all points gathered will be stamped with that time point, we need to eliminate that error in order to reach a better accuracy. The assumption neglects that during the timeframe the points can move either through the movement of the LiDAR sensors (i.e. movement of the car) or through movement of the observed objects on which the points where detected.

Therefore there is an error depending on the actual time the point was measured, the movement of the sensor and the movement of the observed objects/points. That leads to an error for each recorded point in respect to the time and movements. Although this is just one component to influence the confidence ellipsoid of that point, we want to elaborate this exemplary.

$$\epsilon\left(t, \vec{v}_{ego}\left(t, \vec{a}_{ego}(t)\right), \vec{v}_{obj}\left(t, \vec{a}_{obj}(t)\right)\right), with\ t \in [t_{n-1}, t_n]$$

t represents the actual time the point was measured
t_n stands for the time stamps of the frames the sensor recorded.

The velocity itself is being recorded in discrete steps and has to be interpolated to estimate the velocity at the time point of the measurement. Depending on the frequency of the velocity measurements, the acceleration has to be considered for a nonlinear interpolation. The acceleration measurements themselves are also discrete and are being recorded at different time points. While in reality even the different components of velocity and acceleration are measured or derived at different time points and even in different coordinate systems with time dependent transformations. So far we have three different values the error relies upon: velocity, acceleration and time.

Velocity, acceleration and the LiDAR points are being measured at different points in time and at different frequencies. The Velocity is dependent of time and acceleration, while the acceleration is itself dependent of time. Supposing that we are in the 3 dimensional space, we should consider the translational movement and acceleration along all axes and the rotational acceleration and movement around the z-axis i.e. steering input of the driver. It becomes clear that minimizing all errors and uncertainties of the compound raw data in three-dimensional space is a complicated task, which requires a large amount of computation time.

Now the challenge is to fuse the data of different sensors which have different uncertainties that may have probabilistic behaviour, but depend on the time and are being recorded at different frequencies. There are many methods to reduce the uncertainty that has been introduced. However, we do not want to state or explain the methods here, but give a framework that can be used to apply those different methods and to calculate the resulting confidence values.

3 Modular framework for confidence optimization

A common approach for getting a reference system with optimal confidence is to use a sensor set and a specialized software tooling to generate the ground truth for the testing scenario. Those sensors are often very cost-intensive. Most of the time the sensors will be installed on the outer hull of the car and introduce limitations on what can be tested. Some sensors are only suitable for close range scenarios, while others can only be integrated into the vehicles at a restricted maximum speed so that some tests are not possible. To avoid such tradeoffs, the framework that we propose can yield better results relying on not only reference sensors but also post processing. Nevertheless reference sensors will be necessary most of the time to accommodate for limitations of the original setup or for having an independent reference.

The definitions and requirements are needed in advance to build up the reference system. The modularization is needed to allow diverse use cases with different setups. The different modules are grouped by topic. The processing model shows the flexibility and advantages of the modularized approach.

3.1 Definitions and requirements

First of all we need to define what is being observed. The framework shall be able to build a time independent world model with all static entities. It shall give the position of dynamic entities at any point in time throughout the recorded scenario. Every entity shall have a confidence ellipsoid as a metric for their uncertainty.

To be able to test reliably and accurately, the reference needs to be more accurate than the entity that is being tested. In order to determine the required accuracy, we will use the Nyquist theorem. The theorem is originally defined for signal processing to determine the frequency that is needed to reproduce a signal accurately. It states that the frequency has to be at least two times the frequency of the signal to reproduce it reliable. We use that to determine the following criteria:

> *"In order to give a valid statement about the confidence of the tested entity, the confidence values of the reference entity have to be at least twice as precise as their corresponding confidence values in the entity that is being tested"*

Entity is defined as a value or processed result of values, with an uncertainty dependent on time and or space. For example a point, a point cloud, an object, velocity measurements, etc.

This criteria is a simplification in order to have a basic metric, which can be applied automatically to measure the validity of the world model that results from our approach. How the confidence is being compared has been left uncertain on purpose to allow different metrics for different scenarios.

In order to fuse all relevant data into one coordinate system, the temporal and spatial association has to be given.

> *"All data must be timestamped and synchronized in time. The relative position and orientation of all Sensors must be known."*

In this framework the distinction between dynamic and static objects is very important. Therefore, we need to define it in more detail.

> *"An object is static if its geometric center and boundaries do not change throughout the scenario. Inherent movement is allowed."*

A parked car with an opening door is not static, since its boundaries change. A parked car where we can see passengers moving inside is static, since the geometric center and the boundaries do not change. However, it is not always easy to find out if an object is static or not. For example: Is a bicycle rider who moves on the bicycle but not the bicycle itself static or dynamic? Since we use the static objects time independent throughout the scenario problems can arise. For example, if an object is not seen for a longer period of time and has moved in between but is static before and after.

3.2 Modules

In testing and validation there is the need to work with a variety of sensor setups and different processing models. Therefore, all modules need to be able to work with different data sources and different algorithms. The modules are designed specifically to accommodate those needs. Each module is a collection of different algorithms that can be used either alone or together in which case they are combined with e.g. a Kalman filter. We use both classical numerical methods and methods like machine learning algorithms.

- data model and transformations

  The data model is a critical part of the framework. It has to be able to transform all raw data into a format that is usable by all modules and algorithms. Additionally, it must guarantee that all original raw data is linked against the resulting entities. In contrast to an in-car fusion some algorithms will be searching independent of the time for entities. The data models memory footprint is, therefore, much bigger than the original raw data. A seamless integration of a database to store all data and to allow searches is necessary.

- raw data fusion

  Depending on the number of data sources and their sensor type, different approaches will be needed for different scenarios. The main challenge is to automate the fusion depending on the different sources. It is still unclear if an automation is possible for all setups.

- object fusion

  The object fusion is a well-known process as mentioned in section 2.2. The quality of the outcome depends on the quality of the object detection and tracking.

- object detection and tracking

  This module has vast differences in respect to in-car methods. Since in post processing we have the information of all the frames available, we do not rely on the most current frame. The module will first find all static and dynamic objects throughout the scenario in a similar manner to in-car methods. The static objects allow us to get a very precise localization of the ego vehicle and mapping of the surrounding static objects with the help of the localization and mapping module. With that we can enrich our raw data with more accurate information and retransform the original raw data with lower uncertainties especially for points that are far away. This will be repeated until it converges. The problem described is complex and the methods used can hardly be explained in detail here. The same is being done for all dynamic objects afterwards.

- localization and mapping

  This module highly depends on the available sensors. If GPS raw data is present we will use post processing, such as differential GPS with survey markers to correct the accuracy. Afterwards, methods such as Simultaneous Localization and Mapping (SLAM) are used, which include approaches such as loop closure as proposed by [14]. The different methods will be fused with the help of an extended Kalman filter.

- object refinement

  This module will gather time independent information about the objects and stack them together. In the case of LiDAR it will accumulate the raw points and create and 3D point cloud of the model. With the help of the convex hull and camera images color information is added. The result is a very detailed representation of the object.

- Metrics

  The different metrics will calculate confidence values for several aspects of the scenario and provide indications about the quality and accuracy of the resulting model. With the metrics we compare our results against the entities that are to be tested.

3.3 Processing model

To clarify the behavior of the modules we will elaborate the processing model.

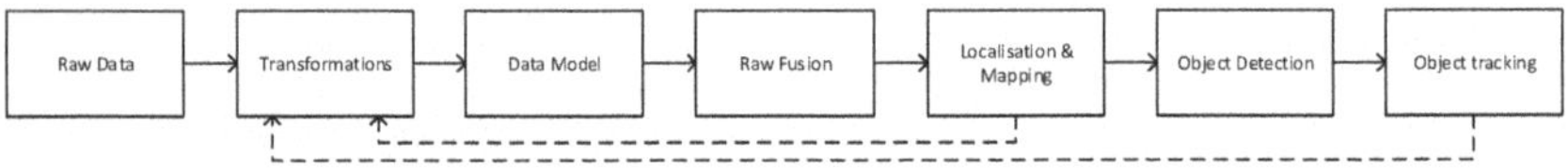

Figure 2 exemplary process model for error reduction in LiDAR data

In this picture we can see how the error reduction can be done within the framework for LiDAR data. The localization and mapping yields a better approximation of the orientation and position of the ego vehicle. That information helps with the initial transformation into the data model. The object detection and tracking yields more accurate information about the orientation and movement of the dynamic objects which again is used for the error reduction explained in (2.3). As outcome we get very accurate data and will then build a very refined scenario.

Based upon the optimized data we build up the model and refine the objects. That model is then used for the actual testing and validation purposes. Bear in mind, that this is just a simplified example for LiDAR. We could also skip the raw fusion and insert an object fusion, or do both. That depends on what kind of data sources are available.

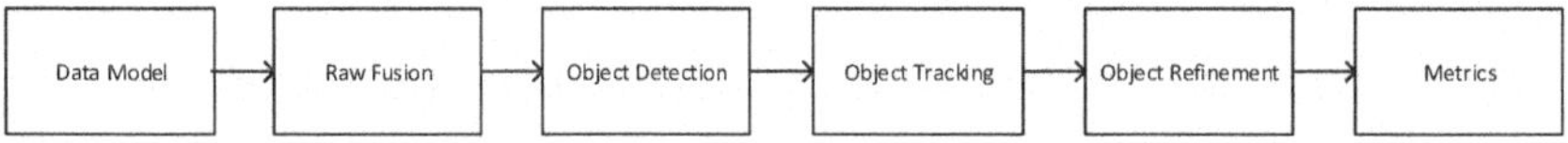

Figure 3 building the refined scenario

The object refinement includes prediction methods, to interpolate the position and posture of dynamic objects outside the visible timeframe. That can be used to overcome problems where the field of view of sensors used for the reference scenario is not covering the same field of view like the sensors that are being tested for example.

4 Conclusion

With the approach we propose we can automate the testing of sensors to a high degree and give detailed results for a variety of use cases. The modularization enables us to change the behaviour of the processing to accommodate new challenges and use cases. Furthermore we can integrate new algorithms seamlessly and fast.

The resulting scenario is very detailed and has the connection to the underlying raw data. One of the next use cases is automatic labelling. Since we calculate the certainty and have the connection to the raw data, we can guarantee the accuracy of labelled raw frames to a certain degree and automate the process.

The framework we propose is by no means perfect and does not account for some of the problems that can arise. To be able to handle arising problems the framework has to be strictly modularized and implemented according to modern programming methods.

Literature

[1] S. Ingle and M. Phute, "Tesla autopilot: semi autonomous driving, an uptick for future autonomy," *International Research Journal of Engineering and Technology,* 2016.

[2] S. L. Poczter and L. M. Jankovic, "The Google Car: driving toward a better future?," *Journal of Business Case Studies,* 2014.

[3] X. Chen, H. Ma, J. Wan, B. Li and T. Xia, "Multi-View 3D Object Detection Network for Autonomous Driving," in *IEEE CVPR*, 2017, p. 3.

[4] F. Castanedo, "A review of data fusion techniques," *The Scientific World Journal* , 2013.

[5] B. Khaleghi, A. Khamis, F. O. Karray and S. N. Razavi, "Multisensor data fusion: A review of the state-of-the-art," *Information fusion,* 2013.

[6] M. Aeberhard and N. Kaempchen, "High-level sensor data fusion architecture for vehicle surround environment perception," in *Proc. 8th Int. Workshop Intell. Transp*, 2011.

[7] J. Zhang, "Multi-source remote sensing data fusion: status and trends," *International Journal of Image and Data Fusion,* pp. 5-24, 2010.

[8] B. Steux, C. Laurgeau, L. Salesse and D. Wautier, "Fade: a vehicle detection and tracking system featuring monocular color vision and radar data fusion," in *Intelligent Vehicle Symposium, 2002. IEEE*, 2002, pp. 632-639.

[9] S. Pietzsch, T. D. Vu, J. Burlet, O. Aycard, T. Hackbarth, N. Appenrodt, J. Dickmann and B. Radig, "Results of a precrash application based on laser scanner and short-range radars," *IEEE Transactions on Intelligent Transportation Systems,* pp. 584-593, 2009.

[10] M. Mahlisch, R. Schweiger, W. Ritter and K. Dietmayer, "Sensorfusion using spatio-temporal aligned video and lidar for improved vehicle detection," *Intelligent Vehicles Symposium,* no. IEEE, 2006.

[11] N. Kaempchen, "Feature level fusion of laser scanner and video data," Doctoral dissertation, Universität Ulm, 2007.

[12] J. Cesic, I. Markovic, I. Cvisic, and I. Petrovic, "Radar and stereo vision fusion for multitarget tracking on the special Euclidean group," *Robotics and Autonomous Systems,* pp. 338-348, 2016.

[13]R. O. Chavez-Garcia and O. Aycard, "Multiple sensor fusion and classification for moving object detection and tracking," *IEEE Transactions on Intelligent Transportation Systems,* pp. 525-534, 2016.

[14]P. K. S. L. Ellon Mendes, "ICP-based pose-graph SLAM," in *International Symposium on Safety, Security and Rescue Robotics (SSRR), pp.195 – 200,* Lausanne, Switzerland, 2016.

Core components of automated driving – algorithms for situation analysis, decision-making, and trajectory planning

Christian Lienke*,
TU Dortmund University,
Institute of Control Theory and Systems Engineering

Manuel Schmidt*,
TU Dortmund University,
Institute of Control Theory and Systems Engineering

Christian Wissing*,
ZF Group, Active and Passive Safety Technology

Dr.-Ing. Martin Keller,
ZF Group, Active and Passive Safety Technology

Dr. Carlo Manna,
ZF Group, Active and Passive Safety Technology

Dr. rer. nat. Till Nattermann,
ZF Group, Active and Passive Safety Technology

Univ.-Prof. Dr.-Ing. Prof. h.c. Dr. h.c. Torsten Bertram,
TU Dortmund University,
Institute of Control Theory and Systems Engineering

* Equal contribution

T. Bertram (Hrsg.), *Automatisiertes Fahren 2019*, Proceedings,
https://doi.org/10.1007/978-3-658-27990-5_17

1 Introduction

Automated driving is a key technology for the future of transportation. There are several motivations to develop automated vehicles. First and foremost, it promises to reduce the number of traffic accidents. Figure 1 shows the accidents recorded by the German police over the past years ([1]) ranging back to 1960. The number of vehicles increased in this time significantly and so did the number of accidents. Most accidents are a result of imperfect and error-prone human drivers. Automated driving helps to reduce these numbers significantly and by that furthermore prevents traffic fatalities. It furthermore would allow disabled persons to regain mobility without being dependent on others and therefore increase the quality of living. Finally, many people perceive driving as a task that causes stress and reduces their productivity. Automated vehicles would allow the passengers to spend their time productively during the journey.

The challenges of creating a safe, comfortable, reliable and trustworthy system for automated driving are manifold. Reducing errors of humans means that human capabilities have to be exceeded. The vehicle has to perceive its environment, reason about future situation evolutions and plan its own motion. Especially the future traffic situation evolution and the planned motion of the vehicle are coupled because of influences between all traffic participants. The uncertainty of the world, the detections of sensor systems and other traffic participants makes automated driving one of the most challenging research projects.

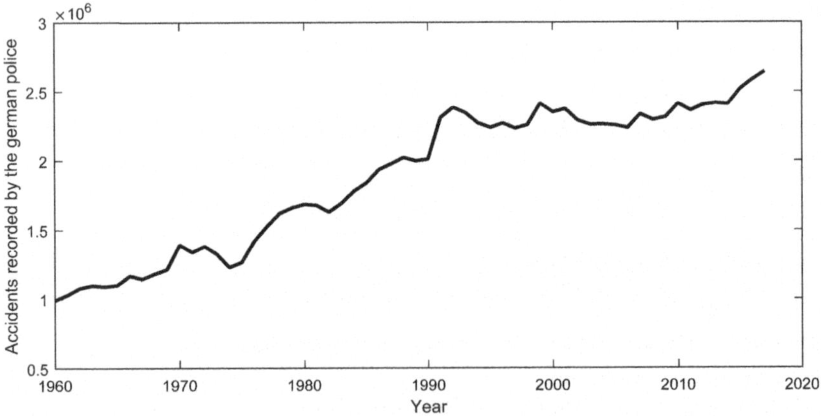

Figure 1: Development of accidents recorded by the German police beginning from the year 1960.

As nowadays series production vehicles only support level two functionality the challenges to reach higher levels of automated driving can among others be seen in the development of algorithms for situation analysis, decision making and trajectory planning. The complexity of these three components in the software architecture of an automated vehicle increases significantly if the system is required to deal with the infinite amount of situations that may arise and without the possibility to fall back on a human driver. To get the most comfortable and safe or in other words best solution for each driving situation it is important to consider the interplay between them.

Existing interdependencies have to be identified and exploited by sharing the gained information at different levels of detail, enhancing the performance of each module and thus contributing to the overall performance of the ego vehicle.

The paper at hand provides an overview of current state of the art algorithms in their respective fields and emphasizes connections and dependencies between the individual components.

The paper is organized as follows: In the next section the functional architecture is described. Then algorithms for situation analysis (cmp. section 3), maneuver planning (cmp. section 4) and trajectory planning (cmp. section 5) are introduced. The interfaces between the algorithms are given in section 6. Results are shown in section 7 and in section 8 the paper is concluded.

2 Functional Architecture

The functional architecture is subdivided into the distinct parts situation analysis, decision making and trajectory planning as well as the interplay between these major components. In the following the main features of the architecture are shortly outlined.

The overall goal of situation analysis is the understanding of the current traffic situation and its future evolution. It builds upon an environmental model, which is generated in the perception layer utilizing various kinds of sensors and the respective fusion algorithm. The model contains information about the states of surrounding traffic participants as well as some representation of the road topology. This also includes traffic rules as speed limits or right of way. The situation analysis result consists of a distribution over future states of all traffic participants, which describes all possible evolutions of the situation and their respective probabilities. To obtain this, various machine learning techniques can be applied. An important aspect one has to consider during prediction is the interaction between single entities as a maneuver of one vehicle can possibly affect maneuvers performed by others (see [2]).

To come to the best future maneuver for the ego vehicle, diverse information has to be merged in the process of decision making. The maneuver is amongst others dependent on demands that are given by route planning as well as on the current and predicted traffic situation, which limits the number of possible actions. Based on the available information the goal of decision making is to determine the best possible maneuver for the ego vehicle. Therefore all possible maneuvers have to be identified in a first step, considering conditions related to comfortable and safe driving on the one side and speed limitations and traffic regulations in general on the other side. Out of all maneuvers the one maneuver, which satisfies all constraints, but also has least negative impact on the other traffic participants, has to be found.

In a next step, the task of trajectory planning is to safely guide the ego vehicle through the traffic additionally regarding the restrictions imposed by higher level modules i.e. the decision making. To consider the changes in the dynamic environment the trajectory planning module is operating in receding horizon manner at high frequency constraining the admissible runtime of the algorithm to just a few milliseconds. This results in a reactive feedback planner that is capable of generating a feasible trajectory under the restrictions given by the decision making module.

3 Situation Analysis

Within the perception part of the functional architecture, various sensor fusion techniques are applied to gather an environmental model of the current traffic situation. However, only the current state is captured. The model does not contain any information on the future evolution of the situation, which is required by the maneuver planning to plan a safe and comfortable driving maneuver. For reasonable decision making a reliable longterm prediction of the behavior of other vehicles in the situation is an important information. To achieve such long-term prediction horizons, the interaction between single participants cannot be neglected in the prediction. The paper at hand applies an interaction-based trajectory prediction approach, which rests upon a Monte Carlo forward simulation of the current traffic situation utilizing probabilistic driver models. Thereby the driver models consider the interaction, and the distribution over future states is approximated by multiple simulation runs. Furthermore, the interaction based prediction can be utilized by the maneuver planning to assess the impact of possible ego vehicle maneuvers by performing the prediction utilizing the maneuver under consideration and rate the predicted impact on other traffic participants.

3.1 State of the art

Following the overview in [3], the prediction of future trajectories for adjacent traffic participants can be divided into physics-based, maneuver-based and interaction-based prediction approaches. The former is only suited for short-term predictions and thus is not in the scope of this paper. Maneuver-based approaches first determine the most likely driving maneuver or intention and estimate a trajectory based on this without considering the interaction with other vehicles. One maneuver-based approach is, for example [4], where the authors combine Dynamic Bayesian Networks for behavior estimation with Bezier curves as trajectory models. Schreier et al. [5] also utilize a Bayesian Network and couple it with individual trajectory distribution models for each maneuver. All these approaches lack an explicit consideration of the interaction between traffic participants, which is mandatory for the coupling with maneuver planning.

In interaction-based trajectory prediction, the interplay between traffic participants is directly modeled, and a more reliable long-term prediction is possible. The authors of [6] combine a model-based interaction-aware intention estimation with a maneuver-based motion prediction based on supervised learning to achieve better results for long-term maneuver prediction. The interaction in between the agents and with the road topology is modeled by a potential field, which guides the agents through the traffic situation in an interaction-based manner. The authors of [7] implement a different approach, where an approximation of future occupancy by other traffic participants is estimated. Uncertainties in prediction are covered by over approximating the future occupancy for other traffic participants, where the interaction limits the reachable area. In [8] the authors apply Inverse Reinforcement Learning to learn a cost function representing human driving behavior and combine it with a vehicle dynamics-based prediction to get the best possible short- and long-term maneuver prediction.

3.2 Interaction aware trajectory prediction

The overall goal in situation analysis is the estimation of the distribution over possible states of all other agents N in the current scene S

$$p\left(\boldsymbol{x}^{(k)}(\tau)\middle|\boldsymbol{x}^{(k)}(t), N, S\right) \text{ with } \tau \in \{t, t+T_p\} \text{ and } k = \{1, 2, \dots, K\},$$

where K is the number of vehicles perceived by the ego vehicle and τ represents the prediction interval from the current timestep with the duration T_p. $\boldsymbol{x}^{(k)}(t)$ denotes the state at time t of agent k. As it can be seen from the equation, the distribution depends on the interaction with other agents as well as the road topology.

To estimate the aforementioned distribution, the approach in [2] is applied. Within a Monte Carlo simulation driver models for longitudinal and lateral maneuvers are utilized to model the interaction between traffic participants and estimate a longitudinal and lateral acceleration for each agent. By applying a basic motion model the future states of all vehicles can be estimated. As the longitudinal motion model the *Intelligent Driver Model* from [9] is utilized. It calculates the longitudinal acceleration of agent k based on the relative distance and velocity of the leading vehicle. For the lateral motion first the lane change model MOBIL (Minimal Overall Braking Induces by Lane change) [10] is used to generate a maneuver decision based on the current situation. Subsequently a maneuver path is sampled from a stochastic process and a lateral acceleration to follow this path is estimated. With the longitudinal and lateral acceleration the state can be updated by the motion model. This sequence is repeated for each agent and for each time step Δt within the prediction duration. To cover various driver types, in the beginning of each Monte Carlo iteration the driver model parameters are drawn from a random distribution such that each agent behaves different in every simulation run.

One major problem when considering interactions between traffic participants is an incomplete environmental model. If a vehicle is not detected by the sensor setup, no interaction with other traffic participants can be modeled. To consider unreasonable maneuvers from the interaction based point of view, a maneuver classification approach utilizing a support vector machine (SVM) is applied to estimate the distribution over the current driving maneuver of agent k. The SVM rests upon object movement related and situational features, where the former describe the current motion of the vehicle relative to its current lane and the later represent features relative to other traffic participants (i.e. distances and velocities). By sampling the initial maneuver of each agent from its respective maneuver distribution the approach can cope with an incomplete environmental model and combines the advantages of an interaction based trajectory prediction with a maneuver based approach.

The overall prediction process is depicted in Figure 2. For further information concerning the mathematical formulation the reader is referred to [2].

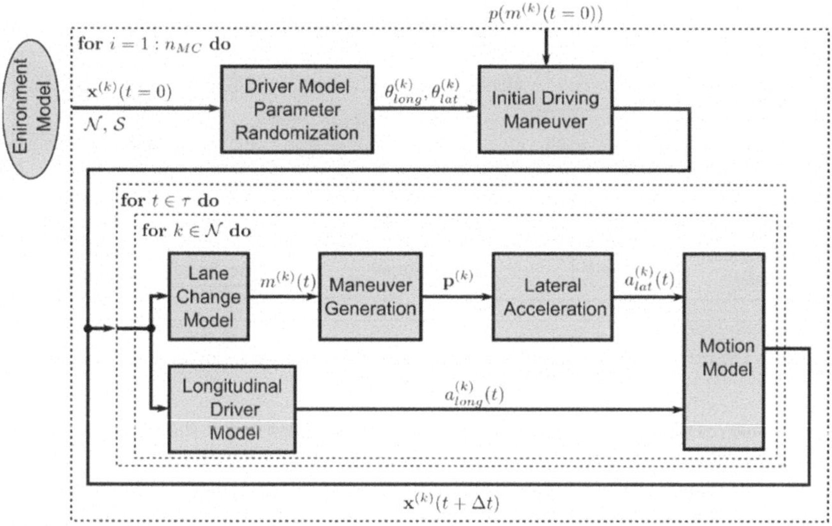

Figure 2: Block diagram of the interaction based trajectory prediction approach in [2].

4 Maneuver Planning

The task of decision making is dependent on various aspects like for example the planned mission and the current traffic situation. Hence, as one part, a maneuver planning module deals with the problem to find suitable maneuvers with respect to the current situation. Figure 3 shows the subdivision of the maneuver planning into the two parts maneuver discovery and maneuver assessment. Its goal is to identify possible maneuvers, calculate a set of candidate trajectories for the ego-vehicle and finally evaluate them. The evaluation includes hard constraint checks concerning driving safety as well as the calculation of vehicle dynamics and traffic interaction features. These quantities are used to assign a cost to each identified maneuver.

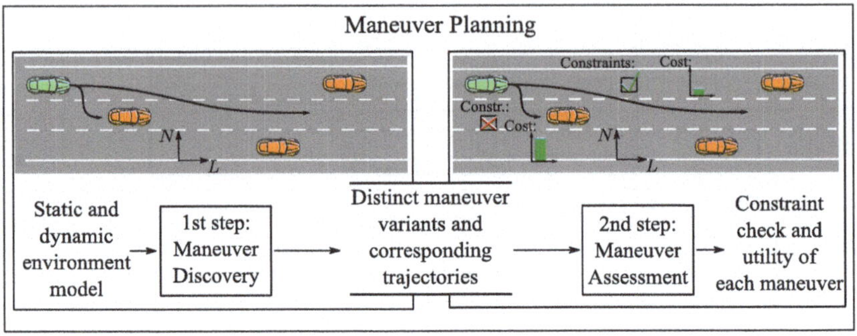

Figure 3: The maneuver planning stage is subdivided into maneuver discovery and assessment.

The maneuver discovery is based on a traffic scene analysis in a spatiotemporal Frenét-Serret coordinate frame (L, N, t). Herein L is the longitudinal, N the lateral coordinate and t corresponds to the time. Possible spatiotemporal lane change corridors that correspond to adjacent spatiotemporal freespace both in the start and target lane are dervided using methods from computational geometry. A detailed description and comparison to state of the art algorithms can be found in [11].

It is crucial that the set of candidate trajectories is constructed in a way to ensure a comfortable ride for the passengers. Lateral trajectories are generated following the approach of [12]. This way, jerk optimal fifth order polynomials are obtained. The approach for the synthesis of longitudinal trajectories is inspired by quadrotor flight [13] and is described next. It differs from the lateral trajectories in the way that constraints are only formulated with respect to positions at three times. These are $L(t_0 = 0\text{ s}) = 0$ m, $L(t_1 = T_{\mathrm{LC,S}})$ and $L(t_2 = T_{\mathrm{LC,E}})$ corresponding to the longitudinal coordinates at the start of a maneuver discovery cycle, the start and end of the lane change.

Lane change maneuver variants are classified into direct and indirect lane changes. The ego-vehicle can directly move toward the target lane in case of direct lane changes whereas it has first to stick on its lane and wait for spatiotemporal free space on both adjacent lanes in case of indirect lane changes. Only one segment $N_1(t)$ is used in case of direct lane changes. The corresponding polynomial aligns the vehicle with the lane centerline and correct orientation of the target lane at the end of the lane change. Two segments are used in case of indirect lane changes. Here, the segment $N_1(t)$ first aligns the ego-vehicle with the centerline and orientation of the start lane and segment $N_2(t)$ finally realizes the transition to the target lane and ensures correct orientation and positioning at the end of the lane change. The calculation of the polynomials is based on the indirect method of optimal control, refer to [12] for the details.

There is no distinction of the trajectory calculation of the longitudinal movement for direct and indirect lane changes. For details of the whole framework, refer to [11]. The approach builds upon the work [13] that presents a robust motion planning scheme for quadrotor flight. This approach was ported for the use in automated driving, especially maneuver planning and is based on the idea to transform a constrained optimization problem into an unconstrained one for robustness reasons. The longitudinal trajectory always consists of two polynomials $L_1(t)$ and $L_2(t)$ linked with continuity constraints. [13] shows that a cost penalizing the squares of derivatives for a single polynomials segment can be formulated using the following matrix equation:

$$J_L(T) = \mathbf{p}_L^{\mathrm{T}} \mathbf{Q}_L(T) \mathbf{p}_L.$$

Herein $\mathbf{Q}_L \in \mathbb{R}^{(n+1)\times(n+1)}$ represents a weight matrix and $\mathbf{p}_L \in \mathbb{R}^{n+1}$ is the vector of polynomial coefficients. The polynomial order $n = 5$ is used. The detailed construction of $\mathbf{Q}_L$ can be found in [14]. For our application, a matrix $\mathbf{Q}_L = 0.7\mathbf{Q}_{L,\mathrm{acc}} + 0.3\mathbf{Q}_{L,\mathrm{jerk}}$ that punishes acceleration and jerk is used. When using two segments, the calculation of a total cost is conducted using:

$$J_{L,\mathrm{total}} = \begin{bmatrix} \mathbf{p}_{L,1} \\ \mathbf{p}_{L,2} \end{bmatrix}^T \begin{bmatrix} \mathbf{Q}_{L,1}(T_1) & \mathbf{0} \\ 0 & \mathbf{Q}_{L,2}(T_2) \end{bmatrix} \begin{bmatrix} \mathbf{p}_{L,1} \\ \mathbf{p}_{L,2} \end{bmatrix}.$$

Next is the formulation of the constraint equation:

$$\mathbf{A}_{L,\mathrm{total}} \begin{bmatrix} \mathbf{p}_{L,1} \\ \mathbf{p}_{L,2} \end{bmatrix} = \begin{bmatrix} \mathbf{d}_{L,1} \\ \mathbf{d}_{L,2} \end{bmatrix}$$

with $\mathbf{A}_{L,\mathrm{total}} \in \mathbb{R}^{2(n+1)\times 2(n+1)}, \mathbf{d}_{L,i} \in \mathbb{R}^{n+1}, i \in \{1,2\}$.

The constraints are placed on position, velocity and acceleration at the segment boundaries. If some are not known in advance, only continuity constraints are imposed. Then the total cost is reformulated:

$$J_{L,\mathrm{total}} = \begin{bmatrix} \mathbf{d}_{L,1} \\ \mathbf{d}_{L,2} \end{bmatrix}^T \mathbf{A}_{L,\mathrm{total}}^{-\mathrm{T}} \mathbf{Q}_{L,\mathrm{total}} \mathbf{A}_{L,\mathrm{total}}^{-1} \begin{bmatrix} \mathbf{d}_{L,1} \\ \mathbf{d}_{L,2} \end{bmatrix}$$

and then reordered for grouping fixed and free derivatives:

$$J_{L,\mathrm{total}} = \begin{bmatrix} \mathbf{d}_{\mathrm{F}} \\ \mathbf{d}_{\mathrm{O}} \end{bmatrix}^T \begin{bmatrix} \mathbf{R}_{\mathrm{FF}} & \mathbf{R}_{\mathrm{FO}} \\ \mathbf{R}_{\mathrm{OF}} & \mathbf{R}_{\mathrm{OO}} \end{bmatrix} \begin{bmatrix} \mathbf{d}_{\mathrm{F}} \\ \mathbf{d}_{\mathrm{O}} \end{bmatrix}.$$

Herein $\mathbf{d}_{\mathrm{F}}$ are the imposed, fixed derivatives on segment boundaries and $\mathbf{d}_{\mathrm{O}}$ are free and can float in order to optimize the total cost of the longitudinal trajectory. The dimensions of matrices $\mathbf{R}_{\mathrm{FF}}$, $\mathbf{R}_{\mathrm{FO}}$, $\mathbf{R}_{\mathrm{OF}}$ and $\mathbf{R}_{\mathrm{OO}}$ depend on the number of fixed constraints. All free derivatives for achieving optimal cost $\mathbf{d}_{\mathrm{O}}^*$ according to $J_{L,\mathrm{total}}$ can be analytically obtained:

$$\frac{\partial J_{L,\text{total}}}{\partial \mathbf{d}_\mathrm{O}} = 0, \qquad \mathbf{d}_\mathrm{O}^* = -\mathbf{R}_\mathrm{OO}^{-1}\mathbf{R}_\mathrm{FO}^\mathrm{T}\mathbf{d}_\mathrm{F}$$

The coefficients of the polynomials are obtained by substitution into the constraint equation. A candidate trajectory set is obtained by combining the longitudinal and lateral trajectories.

Next is the maneuver assessment that is described along Figure 4. Here, first traffic interaction features are calculated in the Frenét-Serret coordinate frame. These include the time gaps, time-to-collisions and distances with respect to surrounding traffic participants. Then a hard constraint check on these quantities is conducted to sort out unsafe trajectories. The remaining trajectories for each maneuver are transferred back into the Cartesian ego-vehicle coordinate frame to obtain lateral and longitudinal vehicle dynamics quantities and form an ego-vehicle cost J_V, refer again to [11] for the details. The user defined number of trajectories with minimum cost J_V is then used for a forward simulation with the corresponding ego-vehicle trajectory injected in the simulation. This is conducted using the simulation [2]. To consider the influence of the ego-vehicle trajectory onto the surrounding traffic participants, an interaction cost J_I is formulated. The underlying idea is to compare kinetic energies of the surrounding traffic participants without the ego-vehicle maneuver and after injecting it. Assuming that every traffic participants wants to maintain its kinetic energy, the difference of these energies can be used for the formulation of J_I. Different weights of traffic participants (compare trucks to motorcycles) are handled via normalization, see [11].

Finally, a user-defined subset of minimum interaction cost trajectories are chosen and the total cost formulated:

$$J_\mathrm{T} = \lambda J_\mathrm{V} + (1 - \lambda) J_\mathrm{I}$$

Herein λ corresponds to a factor to balance the ego-vehicle cost J_V related to ride comfort with the interaction cost J_I.

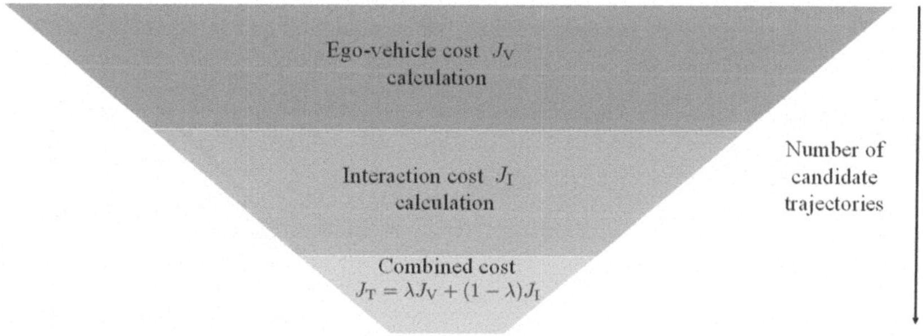

Figure 4: In the maneuver assessment stage, a set of trajectory candidates for each maneuver is evaluated based on vehicle dynamics and interactions with surrounding traffic participants.

Now consider the exemplary traffic scenario in picture a in Figure 5. The goal of the ego-vehicle is to do a lane change to the left target lane. Since the output of the maneuver planning module initializes the local trajectory planner, a total of three maneuver variants arise. The first one corresponds to the ACC (**A**daptive **C**ruise **C**ontrol) and LK (**L**ane **K**eeping) maneuver. There exist two realizations of the lane change to the left lane. The ego-vehicle can do a direct lane change and cut-in before the vehicle that is behind it on the target lane. Another option is to wait until the trailing vehicle on the target lane passes the ego-vehicle. This corresponds to an indirect lane change since the lateral movement is constraint by the time that the trailing vehicle passes the ego-vehicle. The pictures b, c and d show the candidate trajectory sets in green color for each of the maneuvers above mentioned in the spatiotemporal (L, N, t) frame. The blue lines correspond to the lane markings that are perceived by the sensor system of the ego-vehicle. Trajectory predictions of surrounding traffic participants correspond to the orange lines. Both vehicles stick on their respective lanes. Note that the maneuver planning works with an unfixed planning horizon (see [11]) such that the planned trajectories have different durations. A lane change trajectory is handed to the local trajectory planning module only if the among all identified lane change maneuvers, the one with lowest total cost J_T corresponds to a direct lane change. A ACC+LK trajectory is handed over in all other cases.

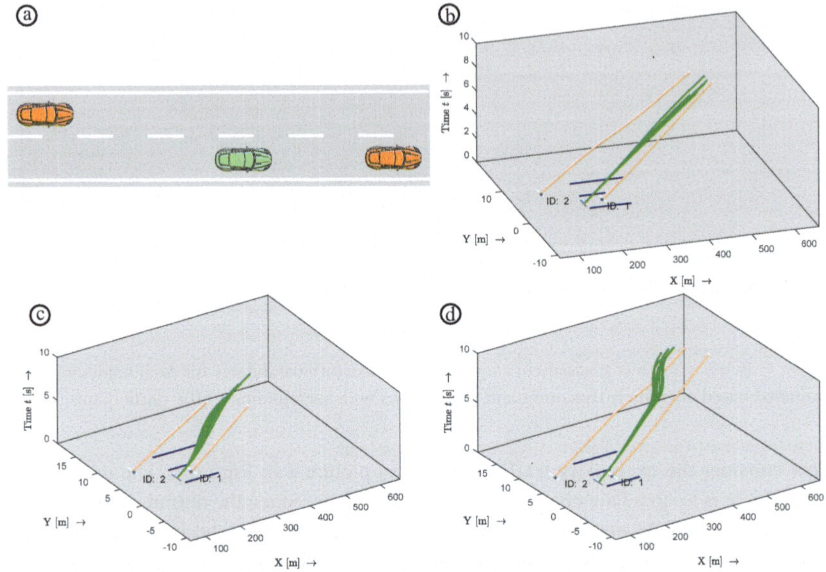

Figure 5: Exemplary traffic scene with the final trajectory sets for the maneuvers Adaptive Cruise Control (ACC) + Lane Keeping (LK), Direct lane change and Indirect lane change. Orange lines correspond to the surrounding traffic participants predicted trajectories and the green lines to the ten trajectories with lowest combined cost.

5 Trajectory Planning

To master the complex traffic automated vehicles have to be capable of planning a trajectory in correspondence to the current situation. Especially in urban scenarios, the application of trajectory planning algorithms is crucial, as the current traffic situation is highly dynamic with high uncertainty in the prediction of the other traffic participants, such that rapid changes in the environment require fast adaption of the planned action of the ego vehicle to the new situation. In the architecture of future vehicles the motion planning module will be responsible to perform a maneuver demanded by the higher level maneuver planning module. The major challenges are the accurate modeling of the vehicle dynamics, real-time capability and exact consideration of restrictions arising from the surrounding environment.

5.1 State of the art

As stated in the survey contributions [15] and [16], the task of motion planning can be accomplished by a broad variety of different approaches. In general the application determines the suitability of a certain approach, as different requirements are obvious for e.g. parking, highway or urban driving. Planning approaches are basically grouped in four categories: Incremental search, graph search, curve interpolation and numerical optimization methods. For normal driving situations incremental search, graph search and optimization methods are the most prominent ones.

The idea of incremental planning is to incrementally build an increasingly finer discretization of the configuration space while attempting to determine if a solution from the initial configuration to the goal region is found. A well-known algorithm in this category is the Rapidly-exploring Random Tree (RRT) [17].

Sampling-based approaches are similar in spirit, as the goal is to approximate a continuous space with a finite number of samples. A compromise between resolution of the configuration space and the number of candidates has to be found, which goes along with the compromise between the quality of the solution and the computational burden. An example of a sampling-based approach for automated driving is given in [18].

Incremental search algorithms and sampling-based methods have in common, that the constraints arising from the environment and the system dynamics of the vehicle are, in comparison to optimization-based approaches, easier to handle. Still the disadvantage of a finite resolution is a problem with respect to the application of automated driving. A more general solution for dynamic environments is the concept of trajectory optimization, which represents a local, continuous method [19]. Similarly, for motion planning this contribution will focus on an optimization-based approach.

5.2 Optimization-based Trajectory Planning

The task of local trajectory planning can be written as a nonlinear program

$$\gamma^*(B, \boldsymbol{\Gamma}_i) = \min_B \sum_{i=1}^{z} \gamma_i(B, \boldsymbol{\Gamma}_i)$$

with respect to equality and inequality constraints

$$\boldsymbol{f}_k(x) = \boldsymbol{0}$$

$$\boldsymbol{h}_k(x) < \boldsymbol{0}$$

where $\boldsymbol{f}_k(x)$ denotes the equality constraints and $\boldsymbol{h}_k(x)$ the inequality constraints. z quadratic functions γ_i contain the objectives $\boldsymbol{o}_i$ weighted with matrix $\boldsymbol{\Gamma}_i$

$$\gamma_i(B, \boldsymbol{\Gamma}_i) = \boldsymbol{o}_i^T(B)\boldsymbol{\Gamma}_i\boldsymbol{o}_i(B).$$

As an underlying trajectory representation a spline is chosen (cmp. [20], [21]). For $\eta - 1$ spline segments, $t \in [t_\iota\ t_{\iota+1}]$ with $\iota = 1,2 \dots \eta$ and interpolation order $\nu = 3, 5, 7, \dots$ spline $\boldsymbol{s}$ is defined as

$$\boldsymbol{s}_\iota(t) = \boldsymbol{c}_{\nu,\iota}(t - t_\iota)^\nu + \boldsymbol{c}_{\nu-1,\iota}(t - t_\iota)^{\nu-1} + \dots + \boldsymbol{c}_{1,\iota}(t - t_\iota) + \boldsymbol{c}_{0,\iota}\,.$$

In correspondance to interpolation order ν conditions for $\boldsymbol{x}_\iota$ as well as continuity constraints $s_\iota(t_{\iota+1}) = s_{\iota+1}(t_{\iota+1})$ the optimal spline coefficients can be determined $\boldsymbol{c}_{0\dots\nu,\iota}$ by optimizing the breakpoints. The breakpoints, denoted with an underscore, are given as

$$B_{TES} = \left\{\underline{\boldsymbol{x}}_1, \underline{\dot{\boldsymbol{x}}}_1, \dots, \underline{\boldsymbol{x}}_1^{(V)}, \underline{t}_1, \dots, \underline{\boldsymbol{x}}_\iota, \underline{\dot{\boldsymbol{x}}}_\iota, \dots, \underline{\boldsymbol{x}}_\iota^{(V)}, \underline{t}_\iota, \dots, \underline{\boldsymbol{x}}_\eta, \underline{\dot{\boldsymbol{x}}}_\eta, \dots, \underline{\boldsymbol{x}}_\eta^{(V)}, \underline{T}_p\right\}.$$

Where derivative $V = (\upsilon - 1)/2$. In this contribution the trajectory planning problem is solved as given in [20]. Herein the nonlinear program is transformed to an unrestricted optimization problem, such that algorithms for nonlinear least squares problems can be applied.

6 Interfaces

A concept in the form of a data flow diagram was realized before all distinct modules were developed. Such concept gives a high-level description of couplings between all involved components. The definition of interfaces is done on the basis of the data flow diagram. All modules are developed and maintained independently, which does not pose any problem because of the aforementioned preparation phase.

Figure 6 shows the information flow between the environment, situation analysis, maneuver planning and local trajectory planning. The green vehicle represents the ego-vehicle and surrounding traffic participants are orange. All features that are used as inputs for the lane change prediction described in section 3 are calculated. The lane change prediction uses three Support Vector Machines (SVM) in form of one-against-all classifiers to output probabilities for the maneuvers lane keeping (LK), lane change left (LCL) and lane change right (LCR) of the observed surrounding traffic participants. These probabilities correspond to prior probabilities of the interaction-aware trajectory prediction approach that conducts Monte Carlo forward simulations of the traffic situation. This results in trajectory predictions with associated probabilities.

The next step consists in the maneuver planning described in section 4. All possible maneuvers of the ego-vehicle are identified during the maneuver discovery stage. An external lane change request determines if possible lane change maneuvers should be identified. This finally results in a set of candidate trajectories that are passed to the maneuver assessment stage. Herein, the trajectories are evaluated concerning vehicle dynamics limits, comfort, safety and benefit for the complete traffic scene. Especially the impact of the ego maneuvers on other traffic participants has to be considered. This is achieved by passing trajectories to evaluate back to the interaction aware trajectory prediction approach and simulate the traffic situation with a given ego trajectory. This way the impact of a specific ego maneuver on the traffic situation can be assessed and maneuvers, which for example result in strong deceleration for other vehicles, can be neglected during the assessment stage. Currently, the best trajectory is chosen and handed to the local trajectory planner, described in section 5, for further processing and finally for execution. The responsibility is therefore transferred to the local trajectory planner after the maneuver was handed over to it. Communication between the maneuver and trajectory planning is currently realized using flags to signalize if the maneuver is currently executed and after finalization if for example the lane change was successful. The loop is closed through the environment because the executed trajectory has a direct influence on the surrounding traffic. This influences the situation analysis and in turn the maneuver and trajectory planning.

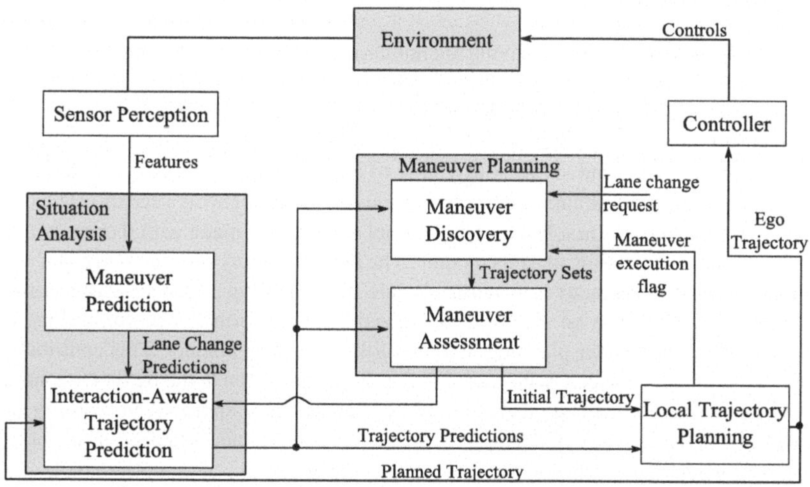

Figure 6: Information flow between all components of the proposed automated driving system.

7 Results

To test the algorithms in simulation, a dynamic environment simulation is used [22]. Camera and radar sensors are modeled such that lane marker information and fused information about obstacle vehicles is provided. A typical highway scenario is chosen. The ego vehicle is driving with a desired speed of 120 km/h and is approaching a slower vehicle (ID:1) driving at a speed of 105 km/h on its own lane. On the left lane another obstacle vehicle (ID:2) starting at $x = 0$ m is driving with a speed of 140 km/h. The maneuver planning initiates a direct lane change to overtake the preceding obstacle vehicle. At this stage the trajectory planning module takes over. The resulting maneuver is shown in Figure 7. After the lane change the maneuver planning module resumes reasoning about the best follow-up maneuver. In a second scenario the obstacle vehicle (ID:2) starts at $x = 50$ m. Thus the gap to make a direct lane change becomes too small and an indirect lane change is performed. The results from this scenario are shown in Figure 8.

8 Conclusion

The challenges of automated driving are numerous, as the task of driving through complex traffic is not trivial. To successfully port the vision of fully automated vehicles to the streets capable algorithms in the field of situation analysis, maneuver planning and trajectory planning are necessary. This contribution introduces state of the art algorithms for each area and shows a way how to couple these modules. The interplay is shown at an example situation which represents a typical highway scenario. The decision, whether to perform a lance change or not can only be made with thorough information from the situation analysis module. The local trajectory planner is dependent on the decision of the maneuver planning module. By providing a maneuver decision the resulting trajectory can be seen as a globalized solution, furthermore diminishing the drawbacks of the shorter planning horizon of the trajectory planner. This contribution shows a promising concept with regard to the challenges faced in the field of automated driving. The results indicate the suitability of the proposed approach in highway scenarios. Even though the complexity increases in rural and urban scenarios, it is expected that the presented concept is extendable to these scenarios.

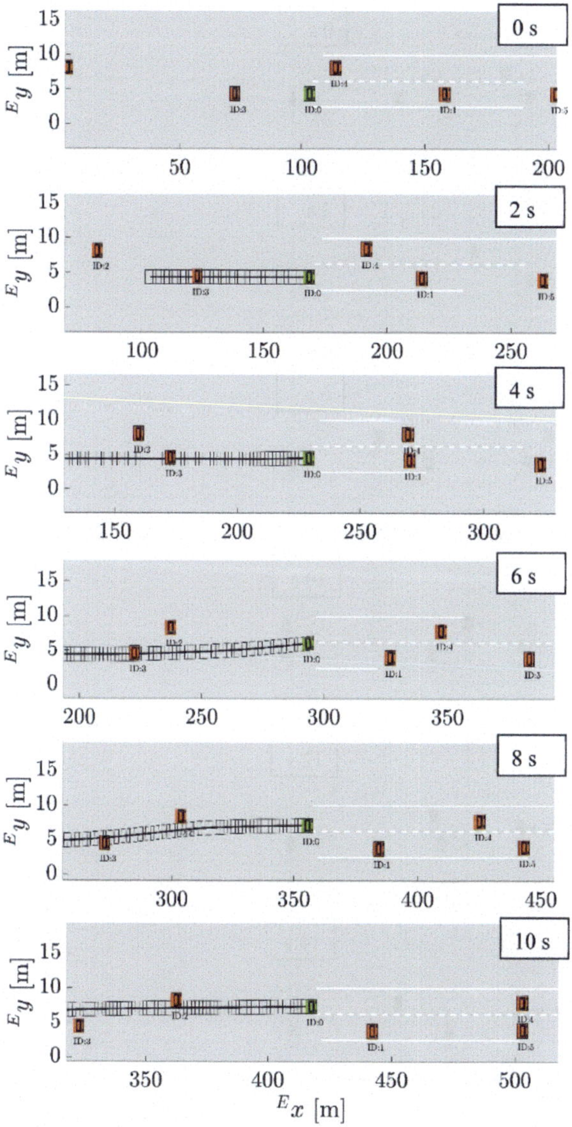

Figure 7: Resulting maneuver in a highway scenario. The ego vehicle (ID:0) performs a lane change to overtake the preceding obstacle vehicle (ID:1) before the approaching faster obstacle vehicle (ID:2).

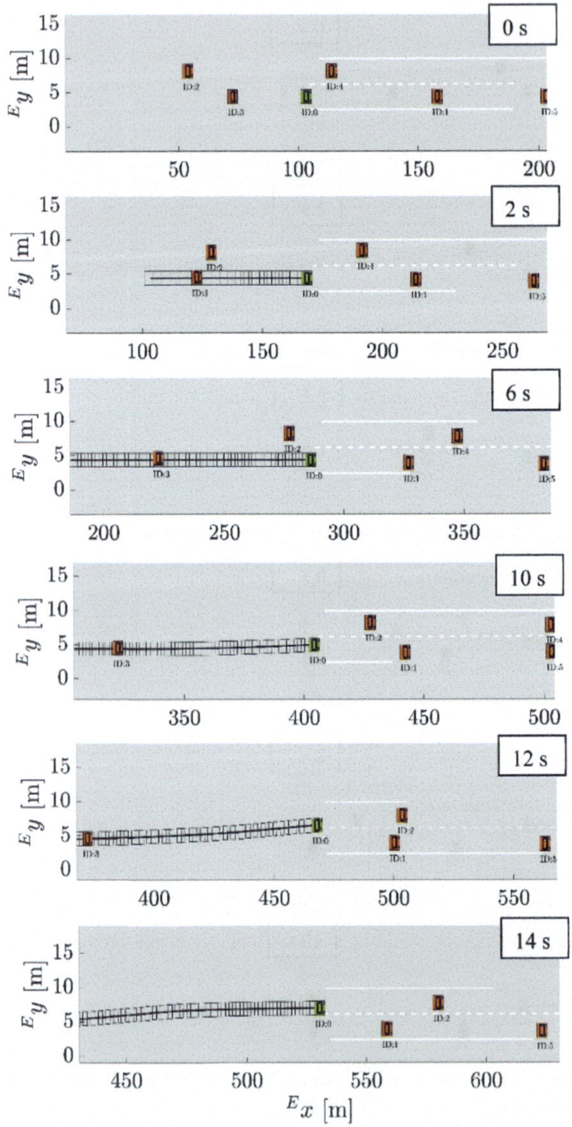

Figure 8: Resulting maneuver in a second highway scenario. The ego vehicle (ID:0) performs a lane change to overtake the preceding obstacle vehicle (ID:1) after the approaching faster obstacle vehicle (ID:2).

References

[1] Polizeilich erfasste Unfälle, Statistisches Bundesamt (Destatis), 2019. In www.destatis.de (Thematische Recherche: Zahlen & Fakten – Wirtschaftsbereiche – Transport & Verkehr – Verkehrsunfälle – Dokumentart: Tabelle). Abrufdatum: 06.02.2019

[2] C. Wissing, T. Nattermann, K. H. Glander and T. Bertram, "Interaction-Aware Long-term Driving Situation Prediction," in *IEEE International Conference on Intelligent Transportation Systems 2018*

[3] S. Lefevre, D. Vasquez, and C. Laugier, "A survey on motion prediction and risk assessment for intelligent vehicles," *ROBOMECH Journal, vol. 1, no. 1, p. 1, Jul 2014.*

[4] T. Gindele, S. Brechtel, and R. Dillmann, "A probabilistic model for estimating driver behaviors and vehicle trajectories in traffic environments," *in 13th International IEEE Conference on Intelligent Transportation Systems, Sept 2010, pp. 1625–1631.*

[5] M. Schreier, V. Willert, and J. Adamy, "An integrated approach to maneuver-based trajectory prediction and criticality assessment in arbitrary road environments," *IEEE Transactions on Intelligent Transportation Systems, vol. 17, no. 10, pp. 2751–2766, Oct 2016.*

[6] M. Bahram, C. Hubmann, A. Lawitzky, M. Aeberhard, and D. Wollherr, "A combined model- and learning-based framework for interaction-aware maneuver prediction," *IEEE Transactions on Intelligent Transportation Systems, vol. 17, no. 6, pp. 1538–1550, June 2016.*

[7] M. Koschi and M. Althoff, "Interaction-aware occupancy prediction of road vehicles," *in 2017 IEEE 20th International Conference on Intelligent Transportation Systems (ITSC), Oct 2017, pp. 1–8.*

[8] D. S. Gonzalez, V. Romero-Cano, J. S. Dibangoye, and C. Laugier, "Interaction-aware driver maneuver inference in highways using realistic driver models," *in 2017 IEEE 20th International Conference on Intelligent Transportation Systems (ITSC), Oct 2017, pp. 1–8.*

[9] M. Treiber, A. Hennecke, and D. Helbing, "Congested traffic states in empirical observations and microscopic simulations," *Phys. Rev. E, vol. 62, pp. 1805–1824, Aug 2000.*

[10] A. Kesting, M. Treiber, and D. Helbing, "General lane-changing model mobil for car-following models," *Transportation Research Record,* vol. 1999, no. 1, pp. 86–94, 2007.

[11] M. Schmidt, C. Manna, J. Braun, C. Wissing, M. Mohamed and T. Bertram, "An Interaction-Aware Lane Change Behavior Planner for Automated Vehicles on Highways based on Polygon Clipping," in *IEEE Robotics and Automation Letters*

[12] M. Werling, J. Ziegler, S. Kammel and S. Thrun, "Optimal Trajectory Generation for Dynamic Street Scenarios in a Frenét Frame," in *IEEE International Conference on Robotics and Automation 2010*

[13] C. Richter, A. Bry and N. Roy, "Polynomial Trajectory Planning for Aggressive Quadrotor Flight in Dense Indoor Environments," in *International Symposium ISRR 2016*

[14] C. Richter, A. Bry and N. Roy, "Polynomial trajectory planning for quadrotor flight," in *IEEE International Conference on Robotics and Automation 2013*

[15] B. Paden, M. Cap, S. Z. Yong, D. Yershov and E. Frazzoli, "A survey of motion planning and control techniques for self-driving urban vehicles," in *IEEE Transactions on Intelligent Vehicles*, vol. 1, no. 1, pp. 33–55, 2016.

[16] D. Gonzalez, J. Perez, V. Milanes, and F. Nashashibi, "A review of motion planning techniques for automated vehicles," in *IEEE Transactions on Intelligent Transportation Systems*, vol. 17, no. 4, pp. 1135–1145, 2016

[17] S. M. LaValle, "Rapidly-exploring random trees: A new tool for path planning," Ph.D. dissertation, Iowa State University, Oct. 1998

[18] M. McNaughton, C. Urmson, J. M. Dolan, and J.-W. Lee, "Motion planning for autonomous driving with a conformal spatiotemporal lattice," in *IEEE International Conference on Robotics and Automation (ICRA)*, pp. 4889–4895, 2011.

[19] J. Ziegler, P. Bender, T. Dang, and C. Stiller, "Trajectory planning for bertha — a local, continuous method," in *IEEE Intelligent Vehicles Symposium (IV)*, pp. 450–457, 2014.

[20] C. Götte, M. Keller, T. Nattermann, C. Haß, K.-H. Glander, and T. Bertram, "Spline-based motion planning for automated driving," in *Proceedings of the 20th IFAC World Congress*, pp. 9444–9449, 2017.

[21] C. Lienke, M. Keller, K.-H. Glander, and T. Bertram, "An ad-hoc sampling-based planner for on-road automated driving," in *IEEE International Conference on Intelligent Transportation Systems (ITSC)*, pp. 2371–2376, 2018.

[22] C. Wissing, T. Nattermann, K.-H. Glander, A. Seewald, and T. Bertram, "Environment simulation for the development, evaluation and verification of underlying algorithms for automated driving," in *AmE 2016 – Automotive meets Electronics; 7th GMM-Symposium*, pp. 9–14, 2016.

Authors

Christian Lienke, TU Dortmund University, Institute of Control Theory and Systems Engineering, 44227 Dortmund, Germany, christian.lienke@tu-dortmund.de

Manuel Schmidt, TU Dortmund University, Institute of Control Theory and Systems Engineering, 44227 Dortmund, Germany, manuel3.schmidt@tu-dortmund.de

Christian Wissing, ZF Group, Active and Passive Safety Technology, 40547 Düsseldorf, Germany, christian.wissing@zf.com

Dr.-Ing. Martin Keller, ZF Group, Active and Passive Safety Technology, 40547 Düsseldorf, Germany, martin.keller1@zf.com

Dr. Carlo Manna, ZF Group, Active and Passive Safety Technology, 40547 Düsseldorf, Germany, carlo.manna@zf.com

Dr. rer. nat. Till Nattermann, ZF Group, Active and Passive Safety Technology, 40547 Düsseldorf, Germany, till.nattermann@zf.com

Univ.-Prof. Dr.-Ing. Prof. h.c. Dr. h.c. Torsten Bertram, TU Dortmund University, Institute of Control Theory and Systems Engineering, 44227 Dortmund, Germany, torsten.bertram@tu-dortmund.de

Do you trust driverless vehicles?

Alfred Eckert
Director Advanced Technology
Systems & Technology
Chassis & Safety Division

Co-Autor: Markus Schneider
Head of Strategy and Innovation

Continental Teves AG & Co. oHG

T. Bertram (Hrsg.), *Automatisiertes Fahren 2019*, Proceedings,
https://doi.org/10.1007/978-3-658-27990-5_18

Summary

Today the level of confidence in many means of transport such as the car, bus, train, and airplane is in general very high, mainly due to the high reliability and safety, which is perceived during using those means of transport. Behind this are continuous improvements and advances in vehicle development and manufacturing, improved infrastructure, and ongoing adjustments of the legal framework worldwide.

Especially regarding cars there have been many activities over the last years to positively influence traffic safety and the safety awareness of road users in the developed countries. The Stop-the-Crash campaign by Global NCAP is rolling this successful procedure out to countries with less rigorous standards.

A residual risk in using the car is socially accepted, while the requirements to the safety standards keep developing in accordance with the respective social environment.

This presentation looks into the question of how much confidence and trust are placed in driverless vehicles today and which measures are necessary to reach an equal level of trust as we know it from proven means of transport. Several market surveys have been conducted and they have delivered varying findings. The assumption is made that the utter basis for any lasting introduction of automated driving – especially driverless vehicles – is provided by vehicle safety and reliability.

It will be discussed why the success of driverless vehicles depends on traffic safety and safety of use, on the general trust of the occupant, and on the comfort of driving, which all together create a feeling of well-being and ease during a trip. Driverless vehicles must perform all tasks of today's drivers (driving tasks, monitoring and communication, including nonverbal) safely and in a reliable manner.

The required technical capabilities are outlined and presented, based on various examples. This includes accident-free driving as the measure of all things for occupants and road users with adapted driving behavior under consideration of the available maneuvering space, also below the legal speed limit in a given situation. This is complemented by fallback-levels and degradation concepts. In addition, the vehicle status must be analyzed to a far greater depth than it is the case today because the driver will not be available as a "sensor". Technical adaptations will be needed to display the changing comfortable horizontal and vertical motion accordingly, and to make transparent to occupants and the environment why the vehicle is reacting in a certain way. And finally, nonverbal and non-technology-related communication between the vehicle and its environment will play a role in creating acceptance of driverless automation.

Introduction

Not too long ago an airplane mysteriously disappeared over the southern part of the Indian Ocean. Others crashed. In the same year when the plane vanished, an airplane was even shot down over the Ukraine. Maritime travel has been confronted with sinking ships since the days of the Titanic and right on to the cruise liner "Costa Concordia", which ran onto ground near Italy and developed a heavy list which cost the lives of 32 people. All of these disasters caused losses and injured many persons. Even rail is not exempt from this if you will consider the catastrophic ICE crash near Eschede.

Fig. 1. *Subjective Impression of Safety – Means of Transportation*

In contrast to the passenger car, these events are singular incidents with many victims, however they become ever rarer. Otherwise no one would be using those means of transport. It is true though that we remember serious plane crashes more easily than the more ordinary car crashes. If you take fatal road accidents, more than one person dies on average per accident, especially because in many crash situations other road users like pedestrians or cyclists are involved.

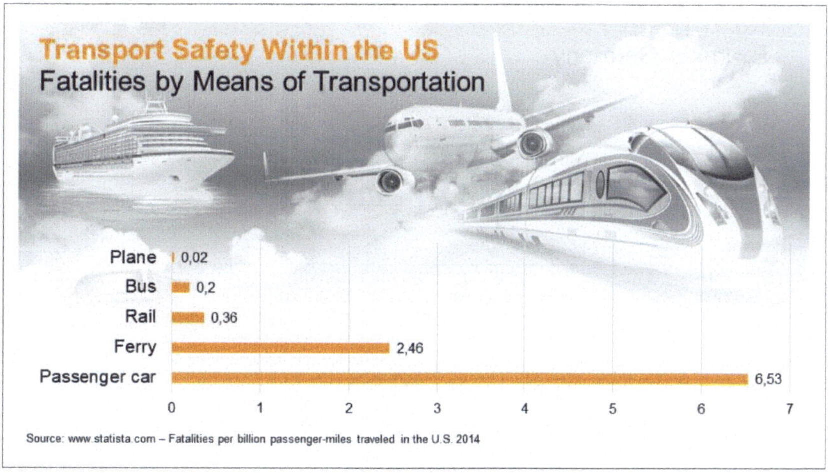

Fig. 2. *Transportation Safety within the US – Fatalities by Means of Transportation*

In general air travel and driving a car are getting safer all the time. Although people have been flying even longer distances over the past years, fewer people die during air travel. Even though there are years when several incidents occur, which causes a statistical peak, the long-term trend goes down (admittedly 2014 was a terrible year but it was still better than many years in the preceding decade).

One can find examples of means of transport which did not advance to a generally used part of mobility even though they may have had a good start. Many reasons can be behind this, some can be technological. but it can also be a simple lack of trust. Consider the Zeppelin which had several catastrophic accidents in its early days (e.g. the Hindenburg at Lakehurst in 1937) and caused many fatalities. Add to this that the Zeppelin has technical disadvantages and you will understand why this technology is now only used for some military purposes and occasionally for tourism (e.g. Zeppelin NT), although the original hydrogen-related risks do not apply any longer.

Another example of a fairly short-lived technology is the paternoster lift which also caused very unpopular accidents with people getting caught in it, getting jammed, being stuck, falling down etc. In the end, new paternoster installations were forbidden in the 70ies.

The importance of trust – or rather fear – could also been seen in the USA in the period following the terrible terrorist attacks of 9/11. After this shocking experience many people avoided air travel and instead used the car even for long distance trips, which would have been covered by the much faster plane prior to 9/11. The example may be a little far-fetched, but it shows which influence a lack of confidence can have on mobility. Eventually official countermeasures (increased security in air travel) were accepted as efficient and the car usage returned to "normal", because the trust in air travel grew back.

New player "Automated Driving"

There is a striking similarity between air traffic, rail and the car: Just like accident prevention has been a constant challenge to all of them, the potential severity of car accidents had to be reduced. This needed to be done although the traffic density in individual regions increased.

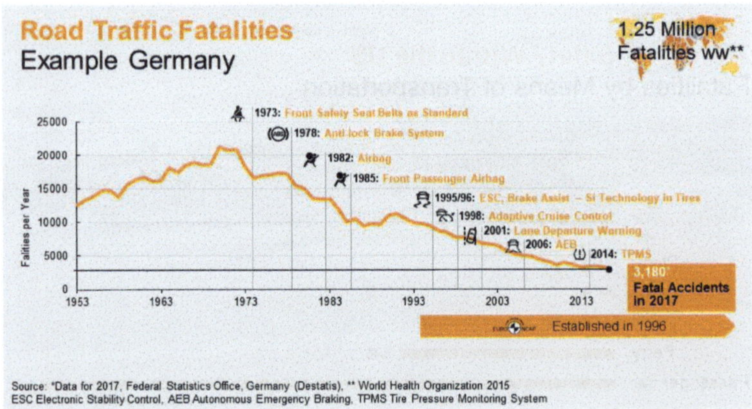

Fig. 3. *Road Traffic Fatalities – Example Germany*

Among the successful measures were legal regulations, and the creation of safety standards applicable to vehicles and the infrastructure. Examples include speed limits, improved properties of the road surface, guard rails, driver assistance systems like ABS, ESC, EBA (Emergency Brake Assist) and technology means like passive safety (e.g. seat belt, airbag). This increasing number of active and passive safety systems was accompanied by a growing awareness about dangers and necessary safety standards. It was supported by road safety education as a precaution against accidents and continued right on to the current "Stop the Crash" campaign and the NCAP activities.

Automated Driving (AD) technologies and vehicles offer the chance to take the driver as a potential (and notorious) cause of accidents out of the equation, to ensure a 100 percent utilization of all installed safety features and to thus further improve the accident statistics of the developed nations, which is a challenge because the numbers of fatalities and injured people have already gone down considerably in those regions. An AD vehicle's "perceived" level of safety needs to be just as high as the perceived level of other vehicles – to ensure an excellent level of acceptance it must actually be higher. One important element of that is not to cause too much public concern about the technology.

The ideal way to achieve this is to further reduce the number of accidents and even more importantly to bring the number of road fatalities as close to zero as possible (= Vision Zero). Only by reaching this goal one can avoid alarming reports which can erode the trust in AD technology before it is even widely available. All efforts to increase road safety by AD technology also need to differentiate between the individual types of AD: Some privately owned vehicles will offer optional AD (Level 3/Level 4), others will not offer any control elements any longer because their design is targeted to an even higher level of automation (L4 to full L5). A fully autonomous vehicle (L5) could fulfill different business purposes and could either transport people or deliver goods depending on the instantaneous demand.

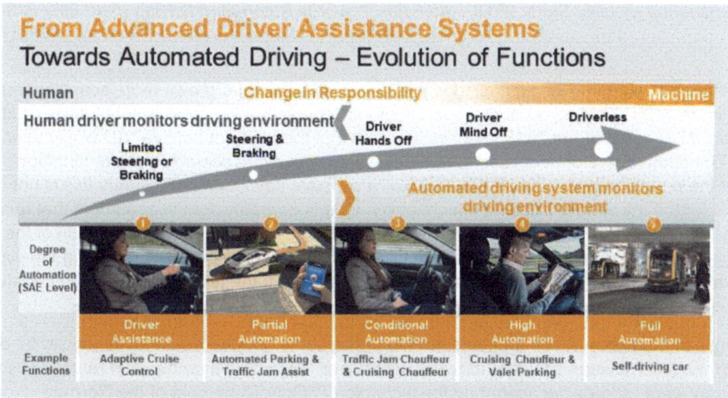

Fig. 4. *From Advanced Driver Assistance Systems Towards Automated Driving – Evolution of Functions*

Obviously, the highest hurdle to give up the control over the car applies to the fully autonomous vehicle without any human driver. As long as there is at least someone monitoring the system (even though that person may not do anything), the sheer presence of this person has a comforting effect. Vehicles with optional AD benefit from this mechanism because the occupants still have a choice.

Various studies, and in particular the recent Continental Mobility Study 2018, have measured amongst others the level of occupants' trust in driverless vehicles and have tried to find out which measures will be needed to reach a level of trust that is comparable to what we know from conventional cars. The insights of those studies, however, vary which is probably owed to the simple fact that nobody has had any experience with AD technology and driverless vehicles. Nevertheless, a few lessons can be learnt from the results of the Continental Mobility Study 2018.

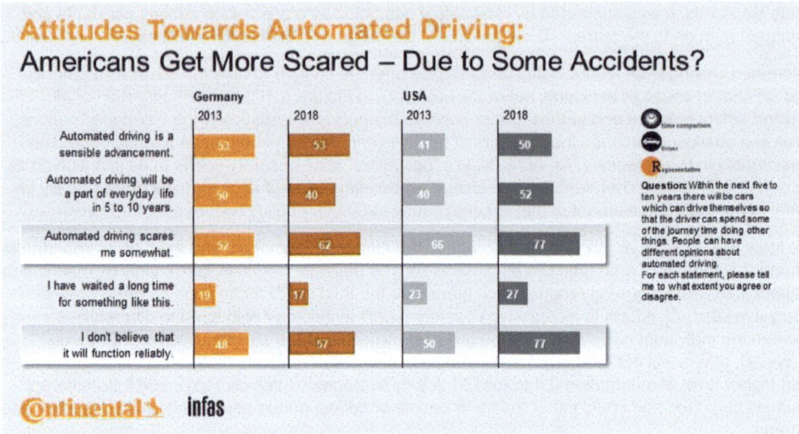

Fig. 5. *Continental Mobility Study 2018 – Attitudes Towards Automated Driving (Germany, USA)*

Probably the most alarming single finding is that drivers in USA have grown more sceptic about AD than they were in 2013 (when an earlier version of the representative Continental Mobility Study was carried out): 77 percent of the 2018 interviewees said that they did not believe "*AD would ever function reliably*". The same share of people said that they were "*somewhat scared by the technology"*. Only 50 percent of the people interviewed considered AD "*a sensible advancement*". Surely this is not a very good basis for the introduction of a new technology. One is probably right in assuming that recent accidents in the USA with automated vehicles (and fatalities!) have tinted the perception of the technology. Although in one case the accident was seemingly the result of over-trusting the system, the general perception appears to be that AD is not (yet) safe. If the technology shall be introduced successfully one cannot simply let this stand as it is.

The situation in Germany is less dramatic, however, the same general tendency is visible: 57 percent of the German interviewees said that they did *not* believe "*AD would ever function reliably*". 62 percent said that they were "*somewhat scared by the technology"*. Only 53 percent of the German interviewees considered AD "*a sensible advancement*". The figures for Japan vary only a little from this and seem to confirm a similar attitude with a slightly bigger acceptance of AD technology.

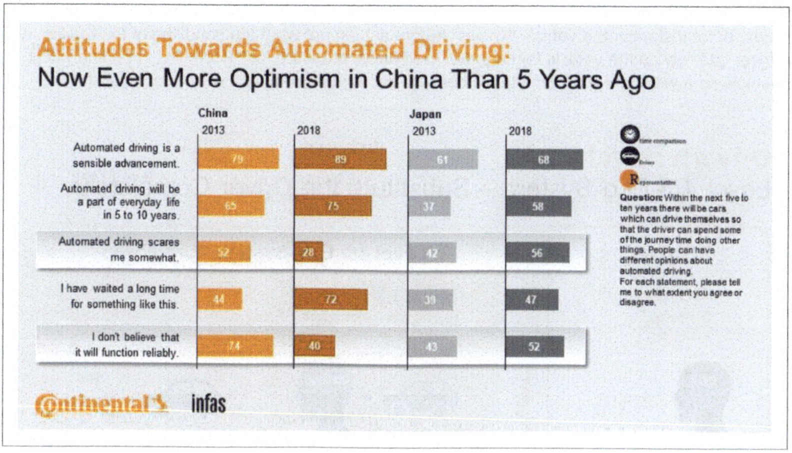

Fig. 6. *Continental Mobility Study 2018 – Attitudes towards automated driving (China, Japan)*

China on the other hand seems to be generally more open to AD technology: Only 40 percent of the Chinese interviewees said that they did *not* believe "*AD would ever function reliably*". Just 28 percent said that they were "*somewhat scared by the technology"*. The vast majority of 89 percent considered AD "*a sensible advancement*".

The figures for the West are quite disillusioning, when we consider that the whole point of AD technology is the added road safety which AD shall bring about! Without acceptance, AD technology will not be widely utilized, and the full safety benefit of the technology will not be realized.

The assumption is made that the utter basis for any lasting introduction of automated driving – especially driverless vehicles – is provided by vehicle safety and reliability.

Admittedly there are other benefits such as the higher efficiency, less emissions and added comfort during higher levels of AD (e.g. being free to engage in other pursuits not related to driving).

Still, if AD technology meets with this level of doubt and skepticism, building trust should be moved to the top of the agenda. The results are only mellowed a little by the observation that younger drivers tend to have a more positive perception of AD. They tend to be more open to it. Still, this skepticism is more than a generation issue. Trust – or rather the lack of it – may easily turn into the main hurdle on the road to AD.

The "compulsory exercise" of AD will therefore be to fully and successfully replace *all* activities of the human driver – and to add the safety of a reliable system which avoids typical human problems such as inattention, drowsiness, emotional stress, or alcohol. AD must function reliably, which means that the driver shall experience a system that can successfully (= no accidents) handle traffic situations. Based on this experience trust will develop over time.

However, there is a second aspect to trust: Advanced Driver Assistance Systems (ADAS) and AD function along a chain of events summarized in the **SensePlanAct** formula. The "compulsory exercise" of the AD car is to get the "**Act**" right. However, to develop trust, many drivers will want to understand the **SensePlan** bit as well! Once a driver understands that the car indeed perceives the environment correctly and fully and draws the right conclusions, trust will grow. This puts great emphasis on the human-machine interaction.

The "voluntary exercise" of AD is to offer the driver benefits, which cannot be found in conventional vehicles. If, for instance, the vehicle "knows" before setting out what the condition of the road is, and if the driver can rely on the vehicle to diagnose itself (tires, pressure, oil, electricity…) he/she will accept the vehicle as a welcome extension to his/her own capabilities and as a welcome support.

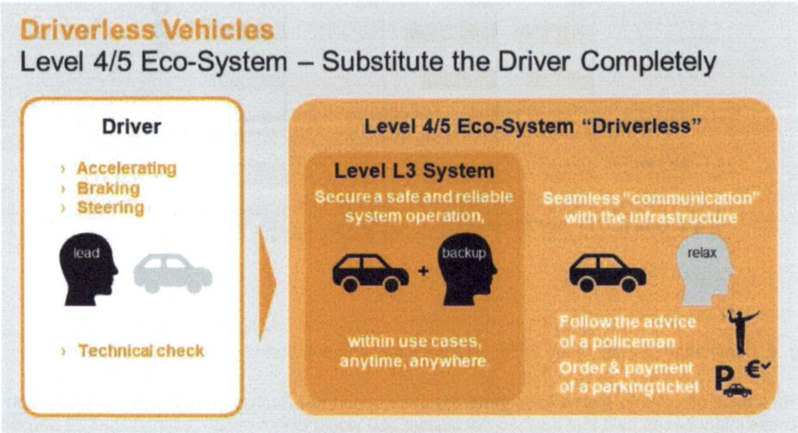

Fig. 7. Driverless Vehicles – Substitute the Driver Completely

During driverless driving (Level 4) the complete communication within the vehicle (without involving the driver) and between the vehicle and its environment needs to be automated and carried out in a bidirectional exchange that can be non-verbal. This needs to be ensured during the entire use case, especially in urban driving, it shall be perspicuous, and safe to support traffic and operational safety.

<u>Some aspects of building trust</u>

The sound basis of driverless driving can only be accident-free driving. It is this capability which generates the required safety of use.

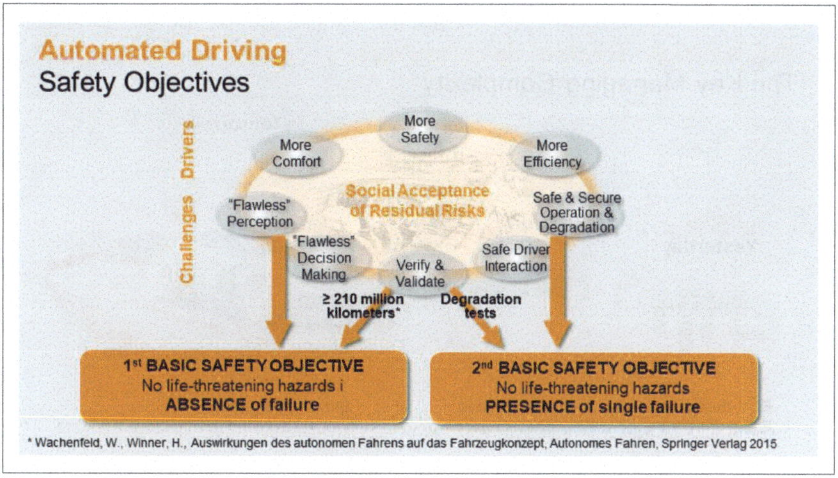

Fig. 8. *Automated Driving - Safety Objectives*

In principle there are two main goals related to safety:

1. Normal driving in traffic needs to be managed without any faulty situations. No critical situation for the occupants or other road users must result from normal driving.
2. A safe reaction of the vehicle in the case of a problem has to be ensured.

Conventional procedures utilize field testing by running a small number of vehicles under the control of an OEM or Tier 1 to ensure compliance with goal No. 1. During this testing phase no problems must arise. Many million miles of road testing provide the basis for AD. In the case of the "Cruising Chauffeur", for instance, more than 200 million kilometers of motorway driving are required to be driven to develop and test the function. This enormous mileage serves just one purpose: To make the AD function perform at least better by one factor than the human driver. This approach is not actionable, though, and will have to be replaced by other procedures such as simulation and silent testing.

Goal No. 2 is addressed through systematic degradation testing.

Nevertheless, there will be a residual risk, which has to be accepted by society – just like it gets accepted with other mobility technologies (flying, rail travel). This appears justified as long as the performance of AD systems is by a factor of X better than the human performance.

To facilitate designing the functional elements of an AD solution and to test them during the system development process against the functional and safety-critical targets, the car needs to have a clear – ideally standardized – and logical structure. One core element of this structure is a powerful architecture which is very different from the current structures and ways of developing. This architecture needs to provide reliable components of safety and a very high availability of the targeted total functionality. Redundancy and fallback-layers in the layout and on the component's level will be an integral part of this new structure.

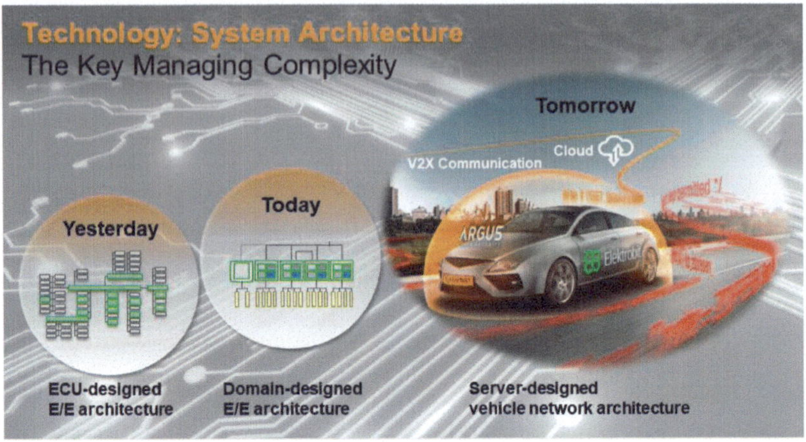

Fig. 9. System Architecture – The Key Managing Complexity

Another key factor is to address the communication with the vehicle environment and the bidirectional influence between car and infrastructure. It will be necessary to give up the old thinking of the car as a system with clear limits. We need to think of the car as a part of an eco-system. One example is the current situation where a driver needs to exit his/her car, for instance, to pay for something. This understanding needs to be widened to – at least temporarily – include functional elements in the infrastructure and to shift functional components to the Cloud (e.g. by distributed computing, software services etc.).

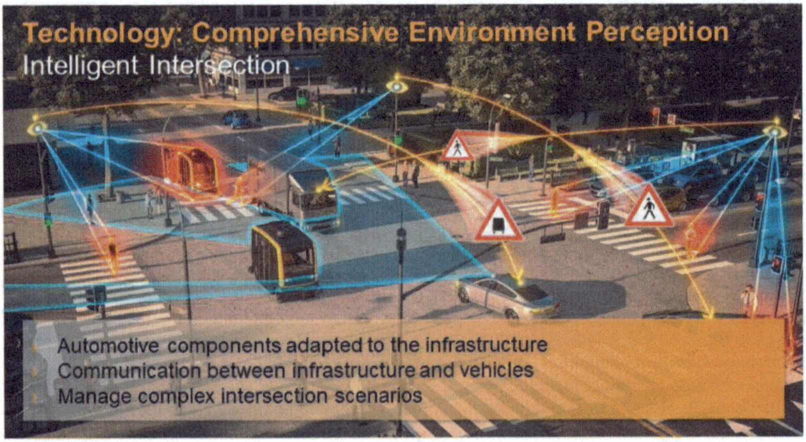

Fig. 10. Comprehensive Environment Perception – Intelligent Intersection

Looking at intersection scenarios, we see the need to have an intelligent infrastructure to widen the perception of the vehicle. This additional information helps to guide the vehicle through complex situations by gaining safety and a more continuous traffic flow.

Gestures coming from other road users need to be processed and answered, if you think of a pedestrian crossing, for instance, or if you consider points of contact to other traffic participants. Also, their acceptance is required to successfully introduce automated driving. Intelligent traffic routing with additional and/or temporary rules must also be factored in.

Fig. 11. *Driverless Mobility – Comprehensive Understanding of the Environment*

In total this asks for an almost perfect understanding of the instantaneous environment to fulfill the use case of an adaptive situation awareness.

The driverless vehicle needs to respect the limits of the available maneuvering space and needs to evaluate this space under safety aspects, e.g., to find the safest trajectory. This capability to adapt and evaluate resembles the ability of a human driver to adapt his/her driving style to a situation where there is either lot of space with excellent visibility conditions versus a situation with narrow lanes and many obstacles which can translate into risks.

Fig. 12. Safety Aspect Road Design

A high level of system availability includes the ability to handle diverse weather conditions, ranging from optimal to quite adverse. Nobody will accept an automation which shuts down as soon as it begins to rain or snow. Robust and ideally anticipatory detection of such boundary conditions plus safe consequences of the AD system will be needed, basically mirroring effective rules such as "reduce speed when wet".

Fig. 13. Road Condition Observer – Including Aquaplaning Risk Recognition

In total the control of the longitudinal, lateral and vertical vehicle movements must ensure that the occupants feel safe and well without developing "motion sickness". Purely technical parameters need to be changed for this purpose, for instance, by shifting the measuring focus of acceleration from the driver seat into the cabin. Anticipatory driving is one of the key factors of success which in turn puts a lot of emphasis on the detection range of the environment sensors but also on information coming from the infrastructure, from or the cloud, or from other vehicles.

Outlook

A sensible strategy to introduce AD will rely on a step-by-step approach offering solid solutions for the applicable use cases and types of AD. Avoiding a "big bang" will help to gradually build trust and/or not to erode it through problems. This is also confirmed by findings in the Continental Mobility Study 2018: The readiness to use automated driving functions and/or assistance systems in specific – and stressful – driving situations, such as passing through the narrow lanes of roadworks, is much bigger than the abstract acceptance of AD as a whole: 67 percent of the German interviewees said that "*automated driving could take over the task of driving for me in monotonous or stressful situations*". This suggests that positive experience with automation during specific use cases may well be the royal road to success.

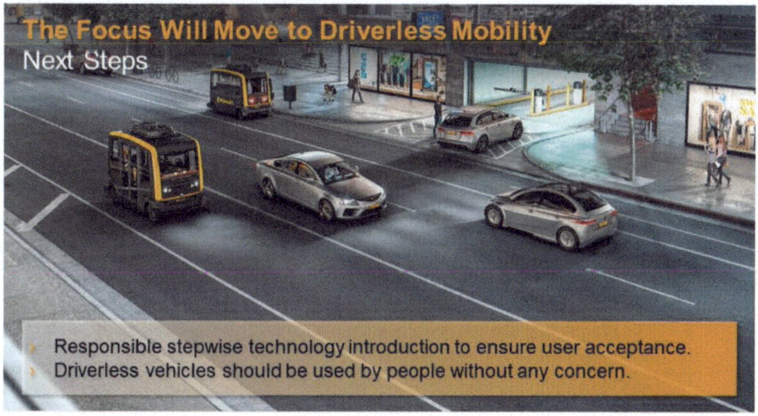

Fig. 14. *The Focus Will Move to Driverless Mobility – Next Steps*

Collecting data will be important to improve products/solutions. However, it must be transparent and socially accepted which data are most suited to support further development. To achieve a maximum level acceptance of data collection drivers will need to know which individual benefits are facilitated in return for providing data. It must be clear who owns the data, how anonymity is ensured and what the added value is that drivers get in return for their cooperation with data collection strategies. This should be a stepped approach where the driver can decide about the scope of data he/she is willing to provide and a clear scope of the stepped services and benefits that can be harvested by the driver in this "barter business". The scope of benefits can range from safety to varying levels of comfort.

Developing AD needs to be a user-centered process not only on the vehicle level but also considering the vehicle environment as other road users need to build trust as well. If this does not work, there is a risk that other road users turn into obstacles for AD acceptance. In terms of acceptance and market success the smartphone shows how an electronic device became massively successful by offering its users an enlarged scope of capabilities and by providing them with a constant access to social media, information, and services. Successful AD will offer drivers a welcome extension of their own capabilities, will take away stress in situations where driving *is not fun*, and will enrich the car beyond a means of transport. As soon as AD is experienced as a useful solution which frees the occupants to enjoy their time in the car AD will turn into a sure-fire success. Trust is essential to get there. And trust is the product of observing, understanding, and accepting what the car does ("My car can handle that"). Once this trust is established, the benefits of using AD will do the rest.

Alfred Eckert

Director Advanced Technology
Systems & Technology

Chassis & Safety Division

Co-Autor: Markus Schneider

Head of Strategy and Innovation

Besucheradresse / Visitor's address:
Continental Teves AG & Co. oHG
Guerickestraße 7, 60488 Frankfurt/Main, Germany

Rechnungsanschrift / Billing address:
Continental Teves AG & Co. oHG
P.O. Box 90 01 20, 60441 Frankfurt am Main, Germany

Tel.: +49-(0)69-7603 3070
Fax: +49-(0)69-7603 88 3070
E-mail: alfred.eckert@continental-corporation.com

Keeping the balance between overload and underload during partly automated driving: relevant secondary tasks

Paula Lassmann[1,*], Matthias Sebastian Fischer[1,*], Hans-Joachim Bieg[2,*], Marcus Jenke[1,*], Florian Reichelt[1,*], Gregory-Jamie Tuezuen[1,*], Thomas Maier[1,*]

1 University of Stuttgart, Institute for Engineering Design and Industrial Design (IKTD), Research and Teaching Department Industrial Design Engineering
* This work was conducted within the project TANGO (Technologie für automatisiertes Fahren nutzergerecht optimiert), which is supported by the Federal Ministry of Economic Affairs and Energy (BMWi) on the basis of a decision by the German Bundestag.
2 Robert Bosch GmbH, Research and Development

T. Bertram (Hrsg.), *Automatisiertes Fahren 2019*, Proceedings,
https://doi.org/10.1007/978-3-658-27990-5_19

Abstract

In partly automated driving the driver has to both monitor the system and be able to take over driving immediately if necessary (SAE Level 2: L2) [SAE16]. The optimal monitoring performance is reached if the driver is neither under- nor overloaded in terms of strain [DeW96]. Advanced driver assistance systems may assist the driver in remaining in this optimal load during potentially long, monotonous automated drives. This is evaluated in a driver simulator study. The study assessed the impact of several secondary tasks (ST), naturalistic as well as standardized, during partly automated driving. 34 participants went through eight test conditions: Two baseline drives (manual driving (L0) and L2 without secondary activity) and six conditions with a L2 automation and different secondary activities. These activities included an auditory n-back task (1- and 2-back) (e.g. [Lor15]) and the surrogate reference task (SuRT) [ISO14198]. Furthermore, an activating task (stretching exercises) as well as a condition in which a video was played were integrated. Subjective strain was measured by using questionnaires (Subjective Experienced Stress: SEA; Stanford Sleepiness Scale: SSS), objective strain in terms of monitoring performance by using a detection task. The results suggest that visual secondary tasks lead to a decrease in drivers' monitoring performance – not necessarily reflecting the drivers' perceived strain. For example, subjectively demanding auditory tasks did not lead to substantial monitoring lapses but may induce positive effects in potentially monotonous automated driving situations.

1 Introduction

Automated driving systems partially release the driver from his driving tasks and thus promise to make driving more pleasant and safer. At the same time, there are increasing empirical indications that partly automated driving (L2) in particular poses risks in terms of the driver's attention: Pure monitoring activities lead to a state of underload after only a few minutes; the monitoring performance decreases significantly [War08]. A countermeasure to prevent such a vigilance collapse could be the implementation of suitable secondary tasks. However, overstraining by a strongly demanding secondary activity can also have a negative influence on the monitoring performance [DeW96].

This study evaluates a comprehensive, systematic selection of secondary tasks during partly automated driving on a highway in regard to their effects on driver attention and strain. This concerns both the selection of modalities (visual, auditory, motoric) and the additional differentiation of the cognitive stress level. This creates the basis for a collection of possible secondary tasks, which an Assistant system for Attention und Activity (AAA) during automated driving can suggest to ensure good monitoring performance. The goal of this study is to collect reliable data for the design of such an assistant system for the project „Technology for automated driving, optimized to the benefit of the user" (short TANGO).

The present article reports only a subset of findings in regard to secondary task demands. The underlying study was more comprehensive and also served as a basis to collect data for an evaluation of automated driver state estimation techniques. Results on this aspect are still preliminary and will be reported in another place.

1.1 Secondary Tasks in Partly Automated Driving

Several studies suggest that drivers' propensity to engage in secondary tasks increases when driving an automated vehicle. For example, drivers were shown to engage more frequently in secondary activities (e.g., cell phone use, watching a video) when SAE Level 1 assistance system such as lane-keeping and automatic cruise control were engaged (e.g. [Mal12, Lla13]). It can be assumed that drivers of L2 vehicles also take up such activities – likely as a countermeasure to the monotonous and unengaging scenario of an automated ride which would require no or only little driver intervention.

Engagement in secondary activities may significantly affect the driver's monitoring capabilities, i.e., to detect and handle automation failures or takeover requests (see studies reviewed by [Gre18]). On the other hand, appropriate engagement may lead to an activation, with potential positive effects on drivers' vigilance [Mil15, Eve09].

For example, physical exercise leads to an improvement of the ability to respond in easy as well as complex tasks (for a review see [Tom03, Cha12]). In addition, it is a meaningful secondary task in terms of health. Therefore, physical exercises are integrated in the study.

Moreover another subject of investigation is the effect of tasks that induce different levels of cognitive stress in a visual as well as an auditory assignment. For assessing the impact of tasks that induce different levels of cognitive stress, standardized tasks were chosen: an n-back task (auditory secondary task) and SuRT (visual secondary task) [ISO14198]. The SuRT, presented on a smartphone, is comparable to cell phone use. With the help of these tasks, a structured investigation of the effect of cognitive stress is possible (for related studies in manual driving see [Con17]). Subsequently we are silhouetted against different studies that used both tasks during partly automated driving, but only in one level of difficulty [Lor15].

1.2 Driver Monitoring Performance

In partly automated driving the driver needs to monitor the road and act when he detects an erroneous behavior of the system. The most reliable result could be achieved by presenting system errors like a slowly progressing lane deviation or an accident ahead.

Unfortunately, the learning effect of participants would lead to distorted results. Therefore an objective measurement method that can measure monitoring performance is needed. One appropriate method to prove reactions to certain stimuli are detection response tasks, which assess hit rate and reaction time. Normally those tasks are seen as a secondary (or third) task and present simple stimuli that do not relate to the main task (e.g. [Ols00, Pat06, Har03]). For a more realistic set-up of the task, studies with visual stimuli that are linked to the simulation environment have been performed. Mainly it was used to show a monitoring performance decrease over time without any secondary tasks. For example drivers had to detect red cars in a platooning condition without a secondary task [Hei17]. A study that focused more on hazardous situation was performed by Greenlee and colleagues [Gre18]. Drivers had to detect cars that disobeyed the rules for priority and would crash into one's own car. Based on the idea of these studies, we decided on a set-up of a monitoring relevant design of a detection response task: the recognition of given speed limit signs (Sign Detection Task: SDT). Hereby we simulate a visual monitoring task. In contrary to the standard detection response task, the SDT is part of the primary task instead of secondary. Via the detection task we want to show if and to which degree a driver is distracted from the monitoring task by a secondary task.

In addition, we are interested in the correlation of subjective strain and the detection response task. Consequently we integrated the Subjective Experienced Strain Scale (SEA) [Eil86]. SEA is the German version of the Rating Scale of Mental Effort (RSME), which is a widely used and reliable method to determine the driver's level of strain (for a review see [DeW16]). We used the Stanford Sleepiness Scale (SSS) for measuring subjective sleepiness and activation of the driver [Hod73].

2 Method

2.1 Apparatus

The test drive took place at the vehicle ergonomics test facility at the research and teaching area Industrial Design Engineering of the University of Stuttgart. The interior is designed as a four-door model with driver and passenger seat as well as back seats (including doors and roof) (Figure 1). The seat and steering wheel position can be individualized by the test person via an electric seat and steering wheel adjustment field. It is a static driving simulation on five monitors (Samsung, 1920 x 1080 pixels, 59") with a 210° field of vision covering, two side mirrors and a rear-view mirror. [Man15]

The HMI includes an instrument cluster (speedometer unit) and a head unit above the center console. A button for the sign detection task (SDT) is positioned in the middle of the steering wheel. As a simulation software, SILAB (WIVW, Version 5.0) has been used.

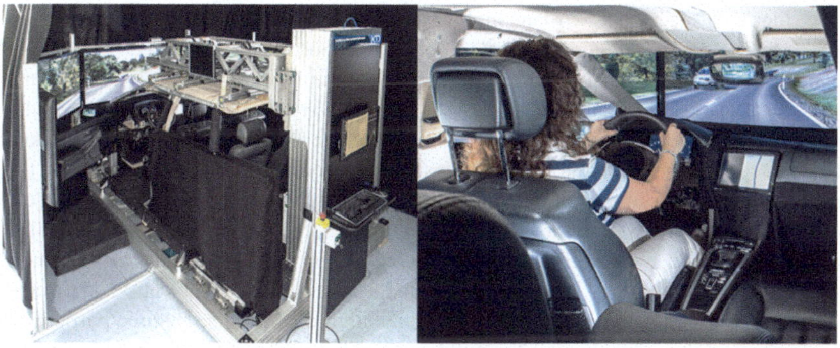

Figure 1: Set-up of the simulator from the outside (left) and inside (right)

2.2 Participants

A total of 34 participants (9 female; 26 %) took part in this study, mainly academic staff members and university students. Their average age was 24.2 ± 2.4 years (aged 18 to 31) with an annual kilometers traveled of 9.738 ± 7.302 km (range 0 to 100,000 km). After a total of two hours of study, the participants received an incentive.

Since this study concentrates on secondary activities, the participants were asked at the beginning about their frequently used media while driving. Most of listen to music while driving manually (85 %). 32 % uses navigation systems. Only a very small number states that they carry out secondary activities that significantly reduce attention (phone call, reading, conversation with co-driver).

2.3 Study Design and procedure

The study was carried out as a within-subject design. Each participant went through eight test conditions during the manipulation ride which are explained in chapter 2.4. The test conditions were randomized.

After a brief introduction of the study and a demographic questionnaire, a familiarization drive was first executed and took about six minutes for the purpose of accustoming the driver to the simulation and to decrease motion sickness. Subsequently, the secondary activities were explained and completed on a trial basis together with the participant. Subjects were told to keep the hands close to the steering wheel in L2 but did not have to hold it (no major effect on reaction time was found between hands on and hands off during take-over [Sch08]). Before the manipulation ride started, the participants answered the Stanford sleepiness scale (SSS).

The rides took place on a two-lane highway with an emergency lane. The start and end of the test sections were always carried out with a short manual drive (L0) of several hundred meters. The transition from manual (L0) to partly automated (L2) driving (and vice versa) was performed automatically by the software. During the entire manipulation ride, the vehicle drove at 100 km/h on the right lane and was occasionally overtaken by other road users. Next, the secondary task was performed if available. In spite of the secondary tasks, the instructor emphasized the priority of the main task: monitoring the road via pressing the SDT button. After seven minutes, the driver was informed that the automation would switch off and was instructed to stop the car manually on the emergency lane. While sitting in the car, participants answered the subjective strain and sleepiness questionnaires. Meanwhile, the simulation continued so that the participant could proceed immediately after the questions had been answered. In Figure 2 the first three test conditions are exemplarily shown. After a total of eight cycles and thus eight test conditions, the manipulation ride was completed.

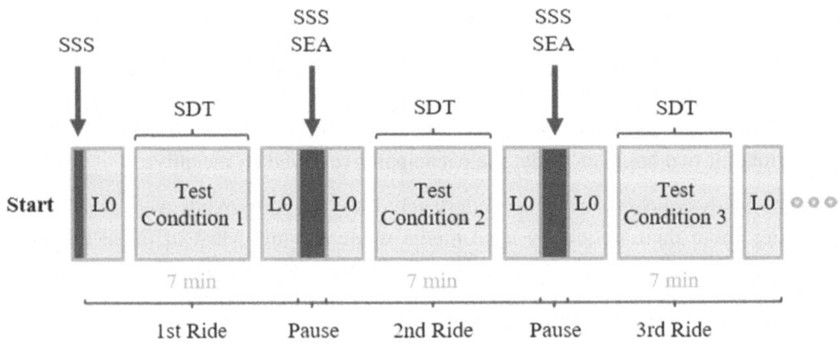

Figure 2: Schematic schedule for the first three secondary tasks (ST) during the manipulation ride

2.4 Independent Variable: Automation Status and Secondary Tasks

The manipulation ride is divided into eight sections or test conditions defined as follows: There was one manual and one partly automated ride used as a reference without performing a secondary activity. Therefore, we refer to them as baseline conditions. The remaining six test conditions took place while the automation (L2) was active, each including a secondary task. Each task required different perception modalities. While some tasks were naturalistic ones, others were standard tasks to ensure comparability and systematic variation of the difficulty, including tasks that are expected to distract the driver from monitoring. The test conditions are shown in Figure 3.

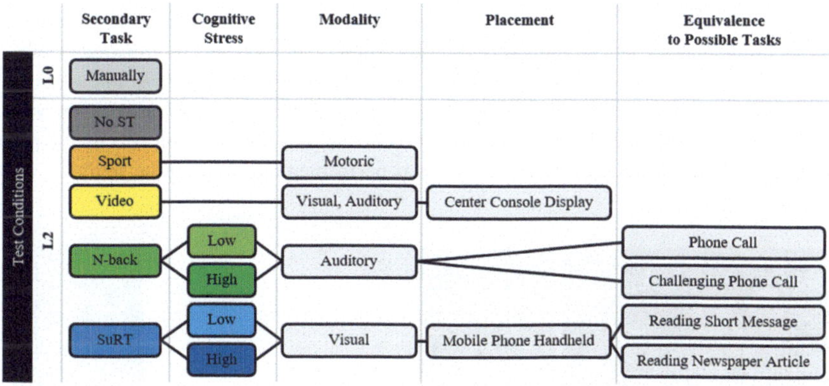

Figure 3: Visualization of the independent variable of the study: test conditions and secondary tasks (ST)

The eight test conditions were the following:

- **Manually:** manual ride without secondary task
- **No ST:** L2 ride without secondary task (ST)
- **Sport:** stretching exercises in the driver's cabin, instructed auditory (primarily motoric)
- **Video:** one brief video documentary on a pre-defined topic of broad appeal (visual and auditory), placement: center console

Cognitive stress was varied within two kinds of tasks (both tasks are explained in detail consecutively):

- **N-back:** 1-back and 2-back (auditory with spoken answer "yes")
- **SuRT:** Surrogate Reference Task (SuRT) [ISO14198] (visual with finger tapping on display), placement: mobile phone, handheld

In the n-back task, a sequence of digits was recited via audio via the PEBL test battery [Mue14] with an Interstimulus Interval of 3000 ms. In case of repetitive digits, the participant was instructed to make a verbal note (spoken answer "yes"). 1-back is the direct repetition of a digit whereas in the 2-back-Task occurs when a repetitive number is mentioned two steps earlier.

The SuRT was presented on a mobile phone (application by German Aerospace Center, DLR; phone: Samsung S3 mini, size: 4.79" x 2.48" screen size: 4"). Distractor circles were set to 30 pixels. In the easier condition the target circle was 40 pixels (Figure 4, left and middle) and in the difficult setting 35 pixels (Figure 4, right). The correct answer was chosen by tapping on the left or the right side on the phone, in which the target circle was located (Figure 4, middle and right). The side was illuminated briefly. Participants were instructed to drop their forearm on an armrest and to hold the phone in either the right or left hand, without switching hand. Furthermore, participants were instructed to perform this task as quickly and correctly as possible, but still with a priority to the SDT. Each secondary activity was performed once during the study.

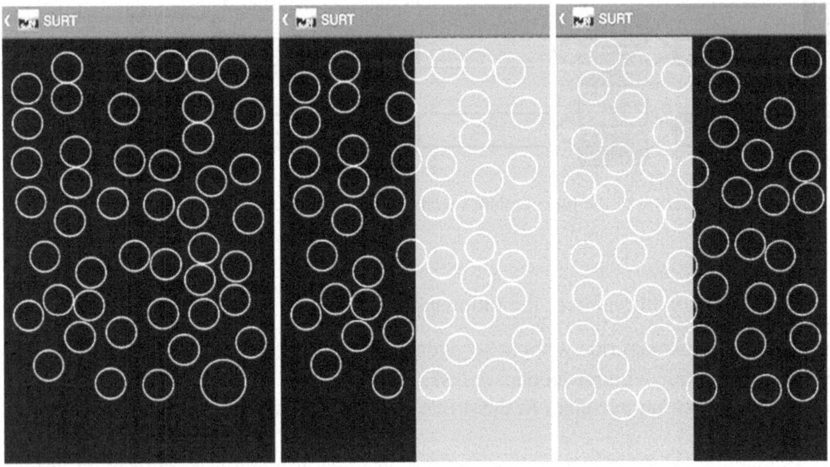

Figure 4: Both configured variations of SuRT – easy (left) with its correct answer (middle) and difficult with its correct answer (right)

2.5 Dependent Variables

An overview of the dependant variables is visualized in Table 1.To simulate the attentional requirements of partly automated driving participants had to press the SDT-button in the middle of the steering wheel whenever 120 km/h speed limit signs appeared (target). During the 7-minute test condition, six 120 signs (targets), three no overtaking signs each and three 130 km/h speed limit signs (non-targets) were randomly displayed at irregular intervals. The minimum distance between two signs was 20 seconds. When the button was pressed, a sound was emitted. This feedback did not signify correct execution of the SDT but was solely meant to inform participants that the button had been triggered correctly. Always two signs appeared at the same time, one on the left side of the passing lane and one to the right of the emergency of the two-lane highway. To prevent a discrepancy in reaction times due to visual acuity, the signs popped up on the sides of the road at a distance of 200 m in front of the car. They remained visible until the vehicle had passed them by 50 m. At a distance of 200 m, the signs could be identified directly by the driver. The button is located in the middle of the steering wheel to prevent effects of left and ride handed.

The participants' subjective strain is assessed by the Subjective Experienced Strain Scale (SEA). SEA is the German version of the Rating Scale of Mental Effort (RSME), which is a method to determine the driver's level of cognitive strain. The SEA is a continious scale with a length of 220 mm. The scale presents anchor points at certain millimeter distances, that help rate the level of strain: "hardly" (20), "slightly" (40), "reasonably" (74), "fairly" (119), "strongly" (154), "very strongly" (187) and "extraordinarily" (204) [Eil86].

To evaluate activating effects of the secondary activity, we used the Stanford Sleepiness Scale (SSS). In order to determine the state of sleepiness, the participant had to assess himself via seven scale points from "active and lively; alert; wide awake" (1) to "almost dreaming; shortly before falling asleep; struggling to stay awake; hopeless" (7) [Hod73].

Table 1: Description of the dependent variables of the study

Stress Measurement	
objective	**subjective**
Sign Detection Task (SDT); Hit Rate (in %), Reaction Time (in ms)	Subjective Experienced Stress (SEA); Value between 0 and 220
	Stanford Sleepiness Scale (SSS); value between 1 and 7

3 Results

3.1 Validation of difficulty level of n-back and SuRT

For validation of the different cognitive load of the n-back tasks, the hit rate of answers between the two was analyzed. In the 1-back test condition subjects answered 99,5 % (SD = 1%) correctly, whereas in the 2-back test condition only 96,5% (SD = 4%; Wilcoxon Signed Rank Test $p < .001$, $z = -3.751$), which shows the an objective difference in the difficulty level.

Owing to an error of the software of the SuRT we could not read out if participants picked the correct side of the display (hit rate). Nevertheless, we can show the difference between the two SuRT conditions of 26 subjects via the time that it took subjects to choose an answer. SuRT easy showed a mean answer time of 1378 ms (SD = 584) while SuRT difficult showed 3302 ms (SD = 983; $t(25) = -11.01$, $p < .001$).

3.2 Subjective Rating of Cognitive Strain Level

Due to the fact that the answers of subjective questionnaires did not follow a Gaussian distribution (KS-Test; $p < 0.05$) all subjective rating data were analyzed using the Friedman Test with post-hoc Dunn-Bonferroni corrections. Six subjects were excluded from the analysis as a result to the fact that they did not fill out the questionnaires properly (n = 28). SEA results show differing ratings for the driving conditions (χ^2 (7) = 108.712, p < .001). Post-hoc tests (Bonferroni significance level p <.006) show the subjective strain in relation to manual driving (Table 2). 1-back, 2-back and SuRT difficult induce more strain than manual driving; SuRT easy is approaching significance. In addition we examined the increase of stain due to the change of the cognitive stress level in the n-back and SuRT test conditions. As to the contrast of the two, either one of the more difficult conditions are indeed perceived as such: SuRT easy – difficult (z = -3.686; p <.001); 1-back – 2-back (z = -4.084; p <.001) (Figure 5).

Table 2: Differences in contrast to manual driving (p < .006) - *) significant distinction

	Manually	No ST	Sport	1-back	2-back	Video	SuRT easy	SuRT difficult
Median	27.5	10.5	25	77.75	133	52.2	59.5	91.25
z		-2.515	-.517	-4.156	-4.532	-1.924	-2.653	-4.181
Significance		.012	.605	.000*	.000*	.054	.008	.000*

Based on the ratings it is possible to divide the driving conditions into three groups of subjective strain based on the SEA interpretation.

- Slight strain: manually, no ST, sport, video
- Reasonable strain: 1-back, SuRT easy
- Fair strain: 2-back, SuRT difficult

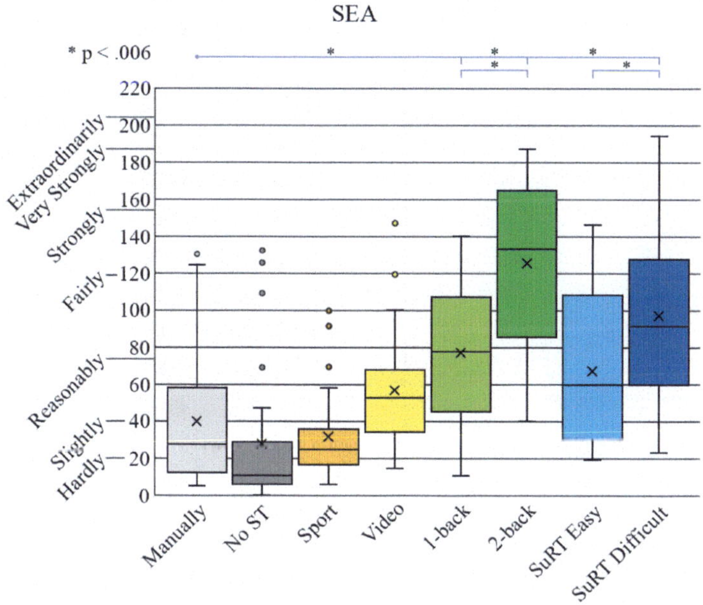

Figure 5: Results of the SEA Scale with significant differences between tasks

3.3 Stanford Sleepiness Scale

The results of the SSS show minor differences. The start value (right before the manipulation ride) and each test condition has a median value of 3, apart from No ST (4) and sport (2) (scale from 1 to 7, where 7 is "asleep").

Results are distinct (χ^2 (8) = 59.912; p <.001). Post-hoc tests show that the start value differs from the two baseline test conditions (manually: z = - 3.343; p = .030; no ST: z = - 4.915; p < .001). Sport contrasts manually (z = 4.052; p = .002), no ST (z = 5.624; p < .001), 1-back (z = -3.410; p = .023) and SuRT easy (z = -3.852; p = .004) as can be seen in Figure 6. Based on these results one can see a tendency that an automated ride without ST leads to dullness whereas sport exercises activate the driver subjectively.

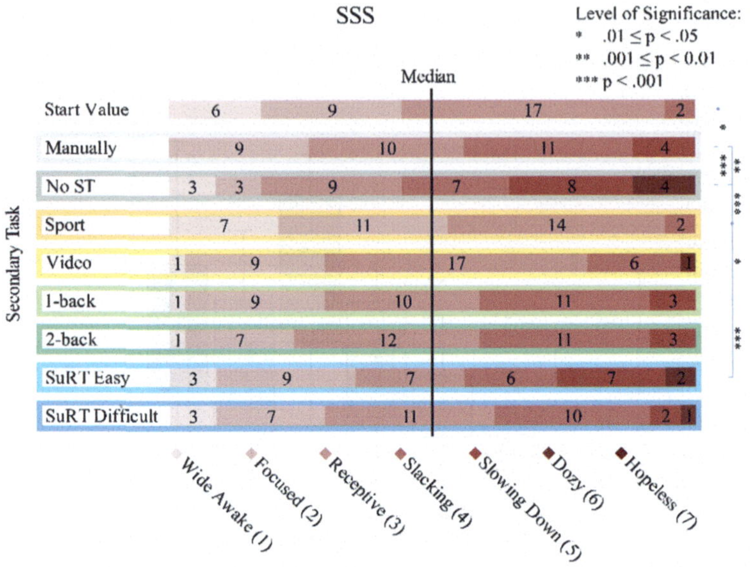

Figure 6: Results of the SSS Scale as absolute frequencies with significant differences between test conditions

3.4 Sign Detection Task

The hit rate of the sign detection task (SDT) exceeded 98 % in all rides. Therefore, an analysis of hit rate differences due to task conditions does not appear to be meaningful. Correspondingly, the analysis will focus on the SDT reaction time (RT) as an indicator of monitoring performance. The appropriateness of secondary tasks in regard to the monitoring activity was assessed by assuming a RT cut-off of three seconds. This time span can be expected for interacting with in-vehicle telematics (2 s eyes-off duration [NHT13], plus 1 s manual reaction time). Tasks that induce longer RTs could potentially be regarded as incompatible with the monitoring requirements of a partly automated vehicle.

The cumulative percentage of RTs of all subjects during one task condition is taken into consideration (Figure 7). For the two baseline rides 94 % (Manually) and 89 % (No ST), RTs lie in an interval between zero and three seconds. RTs of the two (non-visual) n-back conditions (1-back: 96 % and 2-back: 95 %) are similar. A larger proportion of RTs are above the three second threshold in the sport condition (84 %) and in tasks with load on the visual modality (depending on the task load): Video (84 %), SuRT easy (78 %) und SuRT difficult (72 %).

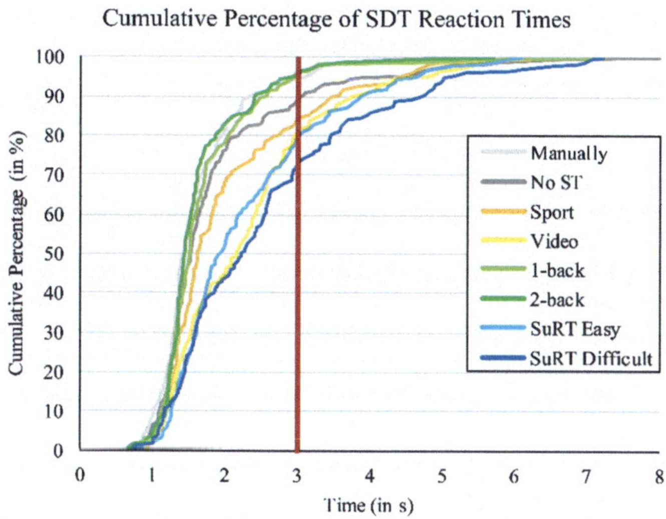

Figure 7: Cumulative Percentage of SDT reaction times

4 Discussion and Conclusion

The results demonstrate a striking difference between objective (SDT) and subjective (SEA) strain evaluations, primarily owing to the different modalities involved in the (visual) SDT and secondary tasks that were tested in the experiment. For example, the SDT showed little sensitivity to auditory tasks with high subjective strain such as the 2-back task, suggesting that an SDT, at least in this form, lacks the qualities of a general device for measuring strain. Yet, for evaluating strain in (automated) driving, a visual SDT seems most appropriate as safety-critical signals tend to occur primarily in this modality (e.g., hazards or lane deviations). Whether this particular visual detection task can be used as a metric for monitoring quality is a subject to discuss. The signs popped up at the side of the road and became bigger as the car was progressing. They were displayed for a long time (7.2 s). In fact, we can compare this task with a suddenly appearing obstacle. Lorenz and colleagues integrated an obstacle in their study and showed that subjects with a cognitive task showed a better monitoring performance than the ones with a visual task [Lor15]. The subjects of the present study were instructed to react to the sign as quickly as possible, but they were not under pressure of a threatening collision. Potentially, RTs would be even shorter in case of a similar event. In contrast a system error in partly automated driving could appear quite differently, e.g. by a slowly progressing lane deviation. In this case, the subjects of [Lor15] showed a better performance while doing a visual task compared to none. The validity of the sign detection task must be investigated further, although we will use it as a basis to rate secondary tasks. Addressing the measurement of objective strain, future studies may consider detection tasks with task-unrelated modalities, e.g., tactile detection tasks, as a promising tool for generally measuring operator strain [Ben12].

In terms of sleepiness and activation we detected effects of secondary tasks during the short duration of one test condition of seven minutes. We found evidence that no ST tends to lead to fatigue whereas sports may activate the driver subjectively. According to [Cha12] the effects of exercising are usually quite small and, as a study on sleep-deprived aviators suggests, do not show long-term effects [Jrl00]. Nevertheless, according to the SSS-results the driver seems to be activated for the time directly after he has been exercising. We could not show an improvement of the ability to respond via the RTs of the SDT. This can be explained by the occupation of both hands in certain exercises.

According to the results of the SDT, drivers' monitoring capabilities were not only unimpaired by the demanding auditory secondary tasks (e.g., n-back) but in comparison to the partial automation baseline without a secondary task, drivers that were engaged in this task even showed increased performance. Therefore, it seems useful to offer occupation, primarily with auditory cognitive tasks. We explain the elevated RT of the sport condition by the fact that in some exercises hands were kept busy as well as the head was turned which led to a distraction of drivers' gaze. Large impairments were caused by visual secondary tasks such as the SuRT tasks or the video task. Designers of in-vehicle information and entertainment systems that are employed in partly automated system should consider this and should seek to design interfaces that promote engagement in non-visual tasks.

On the basis of these results we need to differentiate between subjective overall strain and objective visual strain. High cognitive strain does not necessarily lead to "overload" in terms of the monitoring task. Contrary, it may even promote visual attention. Therefore, a two-dimensional diagram cannot visualize the correlation properly. The dimensions that need to be taken into consideration are the modality and the stress value that lead to different performances dependent on the combination. The findings of this study are in line with the three-dimensional Multiple Resource Model by [Wic02] the effect of multi-tasking. This theory suggests, that two tasks can be executed at the same time without impairment, when they are not dependent on the same modalities. [Wic02] postulates that his model is not a workload model but "predicts performance breakdown in high workload circumstances" which is relevant for monitoring performance in partly automated driving. A demanding visual secondary task can lead to an overload of the visual resources and therefore to monitoring performance impairment, which is demonstrated in this study. A limitation of the model is the fact that underload is not represented. In our case, underload only seems to surface, when there is no secondary task. Therefore we could define underload in partly automated driving as the absence of a secondary task only.

To conclude, the secondary tasks chosen in the study show changes in monitoring performance, dependent on the test condition. Cognitive tasks during partly automated driving lead to a desirable performance comparable to manual driving, whereas no secondary task tends to impair (underload). Overload is exemplified by several visual tasks, such as watching a video or cell phone use. Hence, auditory tasks are relevant in terms of partly automated driving and are planned to be integrated into the Assistant system for Attention und Activity (AAA). Subjective strain assessment does not correlate with the objective measures. That is why the two measures must be regarded separately during studies. In the future we will examine the impact of secondary tasks on a drivers' body behavior with the help of a monitoring system to collect more objective data. In addition, more tests on the SDT should be performed to validate it as a method to measure monitoring performance.

Literature

[Ben12] BENGLER, K.; KOHLMANN, M.; LANGE, C.: Assessment of cognitive workload of in-vehicle systems using a visual peripheral and tactile detection task setting. Work, 2012, 41. Jg., Nr. Supplement 1, S. 4919-4923.

[Cha12] CHANG, Y. K. ; LABBAN, J. D. ; GAPIN, J. I. ; ETNIER, J. L.: The effects of acute exercise on cognitive performance: a meta-analysis. In: Brain research 1453 (2012), S. 87–101

[Con17] CONTI-KUFNER, A. S.: Measuring cognitive task load: An evaluation of the Detection Response Task and its implications for driver distraction assessment. 2017.

[DeW96] DE WAARD, D.: The measurement of drivers' mental workload. Netherlands: Groningen University, Traffic Research Center, 1996.

[DeW16] DE WINTER, J.: Effects of adaptive cruise control and highly automated driving on workload and situation awareness: A review of the empirical evidence. Transportation research part F: traffic psychology and behaviour, 2014, 27. Jg., S. 196-217.

[Eil86] EILERS, K.; NACHREINER, F.; HÄNECKE, K. Entwicklung und Überprüfung einer Skala zur Erfassung subjektiv erlebter Anstrengung: Zeitschrift für Arbeitswissenschaft, 40, 1986, 215-224.

[Eve09] EVERS, C.; Bundesanstalt für Straßenwesen. Auswirkungen von Belastungen und Stress auf das Verkehrsverhalten von Lkw-Fahrern. Heft M, 2009, 204. Jg.

[Gre18] GREENLEE, E. T.; DELUCIA, P. R.; NEWTON, D. C. Driver vigilance in automated vehicles: hazard detection failures are a matter of time. Human factors, 2018, S. 0018720818761711.

[Har03] HARMS, L; PATTEN, C. Peripheral detection as a measure of driver distraction. A study of memory-based versus system-based navigation in a built-up area. Transportation Research Part F: Traffic Psychology and Behaviour, 2003, 6. Jg., Nr. 1, S. 23-36.

[Hei17] HEIKOOP, D. D., et al. Effects of platooning on signal-detection performance, workload, and stress: a driving simulator study. Applied ergonomics, 2017, 60. Jg., S. 116-127.

[Hod73] HODDES, E.: Quantification of sleepiness: a new approach. Psychophysiology, 1973, 10. Jg., Nr. 4, S. 431-436.

[ISO14198] ISO14198. PD ISO/TS 14198:2012 – Road vehicles – Ergonomic aspects of transport information and control systems – Calibration tasks for methods which asses driver demand due to the use of in vehicle systems. BSI, 2012.

[Jrl00] JRLEDUC, P. A. ; CALDWELL, J. A. ; RUYAK, P. S.: The effects of exercise as a countermeasure for fatigue in sleep-deprived aviators. In: Military psychology : the official journal of the Division of Military Psychology, American Psychological Association 12 (2000), Nr. 4, S. 249–266

[Lla13] LLANERAS, R. E.; SALINGER, J. & GREEN, C. A. Human factors issues associated with limited ability autonomous driving systems: Drivers' allocation of visual attention to the forward roadway Proceedings of the 7th International Driving Symposium on Human Factors in Driver Assessment, Training and Vehicle Design, 2013, 92-98

[Lor15] LORENZ, L.; HERGETH, S. Einfluss der Nebenaufgabe auf die Überwachungsleistung beim teilautomatisierten Fahren. Der Fahrer im 21. Jahrhundert, 2015, 21. Jg., 159-173.

[Mil15] MILLER, D.; SUN, A.; JOHNS, M.; IVE, H.; SIRKIN, D.; AICH, S. & JU, W. Distraction becomes engagement in automated driving Proceedings of the Human Factors and Ergonomics Society, 2015, 1676-1680

[Mal12] MALTA, L.; LJUNG, M.; FABER, F.; METZ, B.; SAINT, G.; BENMIMOUN, M. & SCHÄFER, R. euroFOT Deliverable 6.4, Final results: Impact on traffic safety European Large-Scale Field Operational Tests on In-Vehicle Systems, 2012

[Man15] MANDEL, R.; POMIERSKY, P.; MAIER, T.: Der vollvariable Fahrzeug-Ergonomieprüfstand:

Absicherung des digitalen Auslegungsprozesses. In: Binz, H.; Bertsche, B.; Bauer, W.; Roth, D. (Hrsg.): Stuttgarter Symposium für Produktentwicklung 2015, 2015, S. 43.

[Mue14] MUELLER, S. T.; PIPER, B. J. The psychology experiment building language (PEBL) and PEBL test battery. Journal of neuroscience methods, 2014, 222. Jg., S. 250-259.

[NHT13] NHTSA; Visual-Manual NHTSA Driver Distraction Guidelines for In-Vehicle Electronic Devices. Notice of Federal guidelines; Docket No. NHTSA-2010-005. Federal Register, 81, 2013, 78, 24818-24890.

[Ols00] OLSSON, S.; BURNS, P. C. Measuring driver visual distraction with a peripheral detection task. Obtained from August, 2000.

[Pat06] PATTEN, C.: Driver experience and cognitive workload in different traffic environments. Accident Analysis & Prevention, 2006, 38. Jg., Nr. 5, S. 887-894.

[SAE16] SAE. Taxonomy and Definitions for Terms Related to On-Road Motor Vehicle Automated Driving Systems. J3016, 2016.

[Sch08] SCHALLER, T.; SCHIEHLEN, J.; GRADENEGGER, Barbara. Stauassistenz–Unterstützung des Fahrers in der Quer-und Längsführung: Systementwicklung und Kundenakzeptanz. In: 3. Tagung Aktive Sicherheit durch Fahrerassistenz. 2008.

[Tom03] TOMPOROWSKI, P..: Effects of acute bouts of exercise on cognition. In: Acta Psychologica 112 (2003), Nr. 3, S. 297–324.

[War08] WARM, Joel S.; PARASURAMAN, Raja; MATTHEWS, Gerald. Vigilance requires hard mental work and is stressful. Human factors, 2008, 50. Jg., Nr. 3, 433-441.

[Wic02] WICKENS, C. D. Multiple resources and performance prediction. Theoretical issues in ergonomics science, 2002, 3. Jg., Nr. 2, S. 159-177.

Authors

Paula Lassmann[3,*], Matthias Sebastian Fischer[1,*], Hans-Joachim Bieg[4,*], Marcus Jenke[1,*], Florian Reichelt[1,*], Gregory-Jamie Tuezuen[1,*], Thomas Maier[1,*]

3 University of Stuttgart, Institute for Engineering Design and Industrial Design (IKTD), Research and Teaching Department Industrial Design Engineering, Pfaffenwaldring 9, 70569 Stuttgart, Germany, paula.lassmann@iktd.uni-stuttgart.de

* This work was conducted within the project TANGO (Technologie für automatisiertes Fahren nutzergerecht optimiert), which is supported by the Federal Ministry of Economic Affairs and Energy (BMWi) on the basis of a decision by the German Bundestag.

4 Robert Bosch GmbH, Research and Development, 70465 Stuttgart, Germany, hans-joachim.bieg@de.bosch.com

Haptic shared control of electric power steering – a key enabler for driver automation system cooperation

Naoki SHOJI, Research Engineer

Mitsuko YOSHIDA, Research Engineer

Tomohiro NAKADE, Research Engineer

Dr Robert FUCHS, General Manager

System Innovation R&D Department, JTEKT Corporation, Nara, Japan

T. Bertram (Hrsg.), *Automatisiertes Fahren 2019*, Proceedings,
https://doi.org/10.1007/978-3-658-27990-5_20

List of abbreviations and symbols

Abbreviations

AD	Automated driving
ADAS	Advanced driver assistance system
AES	Automated emergency steering
DOF	Degree-of-freedom
EPS	Electric power steering
HMI	Human-machine interaction
HOD	Hands-on detection
LKA	Lane keeping assistance
MD	Manual driving
MRM	Minimum risk manoeuvre
SBW	Steer-by-wire
SW	Steering wheel
TOR	Take-over request

Symbols

T_{assist}	Assist torque
T_d	Driver torque
T_m	Motor torque
T_{tb}	Torsion bar (or torque sensor) torque
θ_{gear}	Pinion angle
θ_m	Motor angle
$\hat{\theta}_{m,ad}$	Estimated automated driving component of motor angle
$\theta_{m,ad,cmd}$	Automated driving command of motor angle
$\theta_{m,ma,cmd}$	Manual and automated driving command of motor angle
$\hat{\theta}_{m,md}$	Estimated manual driving component of motor angle
$\theta_{m,md,cmd}$	Manual driving command of motor angle

Sharing the driving task

Safety in automated vehicles is related to how the driver and the automation system cooperate in the driving task [1]. Intuitive and consistent operation of the vehicle up to SAE automation level 4 through elaborated human-machine interaction is a key enabler for safety, comfort and market acceptance of driving automation [2]. Sharing manual and automated lateral control provides a new driving experience where the driver can steer the vehicle over the automation system without deactivation. Furthermore, coupling the two controls provides the driver with haptic information on where the automation system is heading.

Driving cooperation is often reduced to the authority transfer between manual and automated driving in low level of automation vehicle (SAE level 1 and 2). Override or takeover when using lane keeping assistance is straightforward as the control of the EPS for both manual assistance and ADAS are torque based. Typically, if the driver opposes the LKA torque above a certain threshold, the function is automatically disengaged. Because of the low level of intrusion of such ADAS, the torque threshold is set at a relatively low value and a seamless transfer of authority is achieved. At high level of automation (SAE level 3 and 4) the operation principle of the system changes from temporary intervention to continuous automation [3]. LKA becomes a lane centering, which means that the EPS controls the wheel angle at all time. The angle control in automated driving mode and the assist control in manual driving mode oppose each other. A driver input over the angle control is seen as a disturbance and is rejected. Previous experiences have demonstrated that seamless and robust authority transfer is obtained by a so-called weighted shared control mode during the transition [4]. The proposed approach aims to combine the automated and manual operation into a single, non-exclusive control mode using a weighting function during the transition triggered by the manually induced angle error. While authority transfer is well performed, the full potential of driver–automation system shared control is not exploited.

The literature shows that the concept of shared control and, more generally, of mixing the EPS assistance with the angle control has been addressed in various ways. The levels of implementation in the automated system (Figure 1) and the variety of control objectives demonstrate the wide scope of applications and implication of this concept. However, many publications focus on how to prioritize the driver or the automation system and how to manage conflict, providing limited answers to the technical challenge of developing a control framework enabling both agents to coexist. A non-exhaustive selection of publications is given in the remaining of this paragraph. Benloucif et. al. propose a control that enables driver input by modifying the trajectory of the automated system accordingly [5]. This implies regular rerouting of the trajectory which response is likely not to be optimal (slow bandwidth as indicated in Figure 1).

Cerone et. al. introduce a combined automatic lane-keeping and driver steering with a 2-dof control at the lower level of vehicle motion control [6]. It has the benefit of offering faster response but the goal is to enable an exclusive manual or automated control mode. Abbink et. al. address the problem of shared control at the actuator level optimizing the response time [7]. The control is based on knowledge of neuromuscular control, which limits its market acceptance due to low robustness. More recently, Klesing et. al. presented a work on operating a steering system in a dual-model vehicle [8]. A shared control scheme is proposed for a steer-by-wire system where the two driving modes coexist electronically only. At first the driver performance in following the trajectory of the automated system is evaluated prior to enabling the driver input as the command of the wheel angle control.

This paper presents a control framework for EPS where both manual and automated driving can coexist and which enables cooperative driving. As recommended in [1], shared control is implemented at the operational level. Technically, the classical assist control of EPS active in manual driving and the angle control used for the lane centering are mixed without internal modification in a two degree-of-freedom control architecture. It is a cost-effective solution because of relying solely on the available sensors of the EPS. From the operational point of view, any manual intervention is intuitive as the driver only needs to hold, steer and release the steering wheel. No switch or button activation is required reducing the driver workload while improving comfort and safety. The driver naturally steers the vehicle over the automated actuation. Furthermore, this control provides haptic information on where the automation system is heading when manually steering.

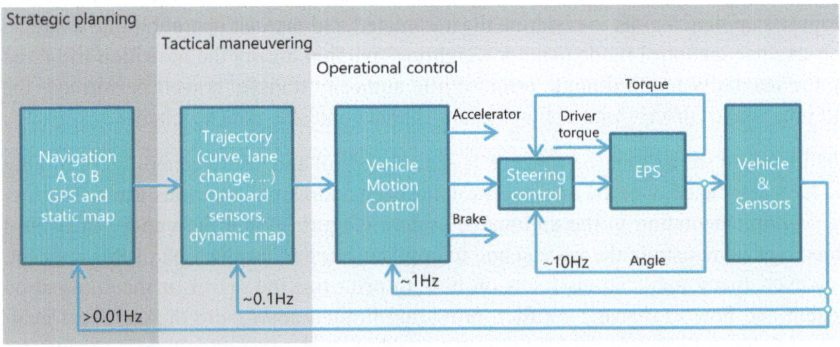

Figure 1: Schematic structure of the automated driving control. The steering control only is detailed. Accelerator and brake controls are not illustrated for clarity reason.

An example of scenario where haptic shared control provides benefits is the passing of a slow moving car on a normal road (with a single lane for each traffic direction). The vehicle is highly automated. The AD system is activated and the driver is released from the driving task. The car in front, moving at a slower pace, makes the cruise control to reduce the speed accordingly. When the driver wants to overtake, naturally, he holds the steering wheel, proceeds to the lane change and return back to the right lane after passing the vehicle. While going straight on the left lane (against traffic), the driver feels a small torque applied to the right direction indicating the location of the planned automated trajectory. That small torque was cancelled for safety by the cooperative control, which set the steering mode to manual while both vehicles are side-by-side. After completing the overtake manoeuvre, the driver is free from holding the steering wheel. No switch or button manipulation was required.

The paper introduces the concept of haptic shared control applied to an electric power steering. The control problem is formulated prior to presenting the proposed control framework, its performance and haptic feedback. The concept of shared control is then extended for consistent implementation with the cooperative control. Examples of applications are given before concluding the paper.

Control problem formulation

The definitions of shared and cooperative controls used in this paper are first stated before formulating the considered problem. From a system (automated driving) perspective, the operational or actuator control gets a command from the higher tactical (guidance) and strategical (navigation) levels (Figure 1). Shared control is implemented at the operational level or in the EPS. Human-machine cooperation is considered at the tactical or strategical level [1]. While shared control provides a framework where both agents can coexist, cooperative control makes the decision on how steering authority is allocated (Figure 2).

Figure 2: Schematic representation of shared control. It provides an operating environment where AD, MD and any level of blending can be set. Authority command is given, for example, from the output of the cooperative control.

The problem consists in developing a closed-loop control system able to track the trajectory of the automated system and to allow any superposed intervention from the driver to modify the vehicle lateral dynamics. It is assumed that both angle and assist controls are available. Technically, the classical assist control of the EPS used in manual driving and the angle control operated in automated driving are combined without internal modification in a closed-loop control framework. Additionally, haptic feedback should be provided to the driver while steering for safety reason.

The following specifications must be satisfied with respect to the shared control problem under consideration:

- Prior to any driver steering action, the vehicle tracks the trajectory planned by the automated system in a continuous manner (no signal interruption).
- When the driver initiates a manoeuvre, the angle control remains active but the vehicle trajectory is modified seamlessly (no external steering wheel jerk, i.e. jerk that is caused by other source than the driver).
- While steering, the driver feels a torque indicating, for example, where the AD is heading (haptic feedback for safety).
- If the driver releases his hands from the steering wheel when the vehicle is on the AD trajectory (i.e. the error of the angle control is null), the vehicle lateral motion is smooth (no external steering wheel jerk).
- If the driver releases his hands from the wheel when the vehicle is away from the AD trajectory, the vehicle goes back automatically to that trajectory (safe vehicle return should be considered at tactical and strategical levels).

Haptic shared control of EPS

This section describes at first the two configurations proposed for the haptic shared control of EPS. Then, performance and haptic feedback are discussed along with test data.

Two configurations

Two control approaches are possible for the assistance of a steering system. The classical assistance, applied to most passenger cars, takes the form of a torque boost. The driver torque, mechanically applied to the column is assisted by a motor torque generated, in its simplest form, by an amplification map. A less common form of assistance is when the EPS is angle controlled. The measured driver torque and its amplification are converted into an angle command to be tracked by the EPS. This configuration is applied, for example, in steer-by-wire systems where the rack is controlled in position from a command generated at the mechanically disconnected steering wheel.

Consequently, two configurations of the haptic shared control are considered. They are named *TA* and *AA* shared control respectively. As a convention, the first letter indicates the control type of the manual assistance (T: Torque control or A: Angle control) and the second letter the control type in automated steering mode (A: angle control). Note that a torque control of automated steering is also possible but not considered in this work. AA and TA configurations will be described in the remaining part of this section.

Control structures

AA and TA shared control configurations are illustrated in Figure 3. As stated in the control problem formulation, the assistance and the angle control are implemented in their original form for both configurations. They are not modified from their independent development.

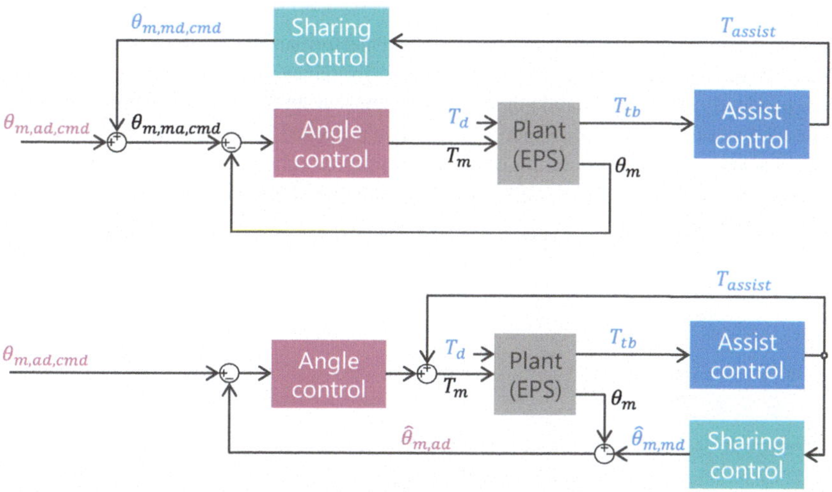

Figure 3: AA shared control (top), TA shared control (bottom).

In both configurations, a sharing control is used to enable the two controls to work together on the plant. Basically, it is composed of a model of the EPS (Figure 4), which computes the angular displacement generated manually from the measured driver input. In the AA configuration, this angular component or MD command is then added to the AD command as input to be tracked by the angle controller. In this way, both AD and MD commands are tracked together by the EPS. In the TA configuration, this MD component is subtracted to the measured angular displacement so that the angle control sees only the remaining AD component. Here, the sharing control plays the dual role of hiding the MD component of the steering displacement to the angle control and, consequently, enabling lateral motion from a driver input.

Figure 5 gives an example of shared control data measured on a driving simulator equipped with an EPS. The AD trajectory follows the path of a winding road. Between about eight and fourteen seconds of simulation, the driver steers the vehicle away from the roadwork. The steering angular displacement shows that the AD trajectory command is well tracked before and after the driver manoeuvre. During the driver intervention, the manual action is superposed to the AD trajectory according to the stated objective. Furthermore, the EPS applies a small torque directed toward the AD trajectory as haptic information.

Performance

Conceptually, both configurations have similar performances. As the assist and angle controls are used in their original form, their performances are not degraded. Based on independently stable assist and angle controls, the sharing control can be tuned to guarantee the stability of the combined system. The proposed controls offer performances robust across the whole operating envelop because not relying on any state machine or discontinuous function. No disturbing jerk is generated on the steering wheel during a driver action.

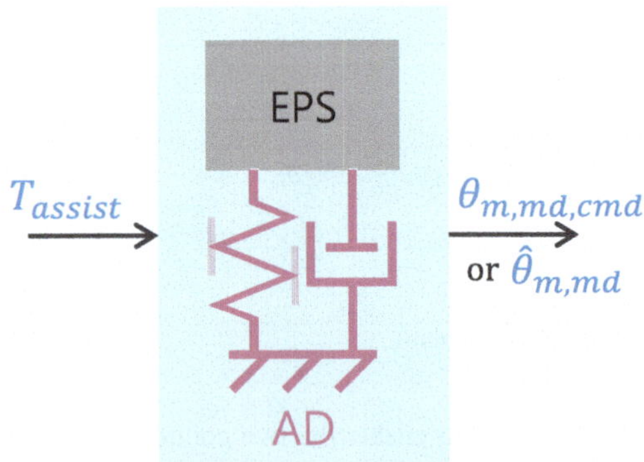

Figure 4: Example of structure for the sharing controller.

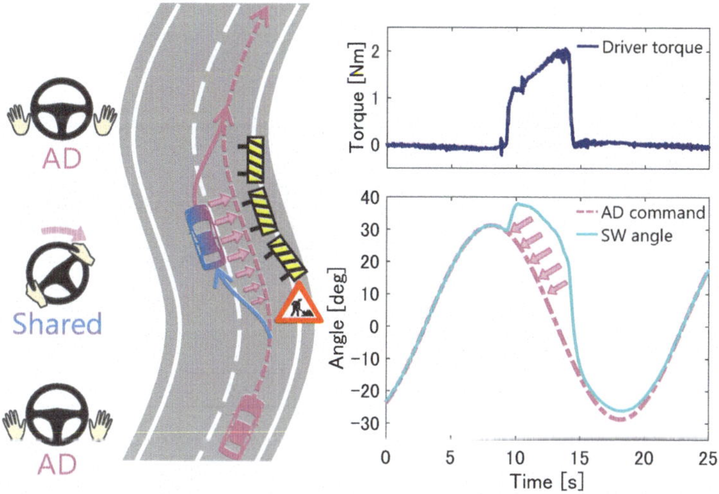

Figure 5: Concept validation made on a driving simulator equipped with an EPS. The AD trajectory follows the path of a winding road. At the location of the roadwork, the driver adjusts the position of the vehicle. During this manual intervention, he feels a small torque in the direction of the AD trajectory.

Haptic feedback

One of the objectives of the sharing control is to provide a haptic feedback to the driver while manoeuvring. The content of this feedback can be designed, for example, to indicate the automation system heading direction. This function is realized with the load applied to the EPS model of the sharing controller (Figure 4). In its simplest form, this load is represented with a spring and damper connecting the EPS inertia to a ground. Instead of using an absolute ground, a relative reference corresponding to the AD command is preferred. With this, the reaction torque felt by the driver is related to the manually induced deviation from the AD trajectory. Furthermore, because of the spring effect, if the driver releases his hands, the steering wheel returns to the AD trajectory automatically. This provides significant benefits in terms of safety, first for the driver, as the haptic feedback informs him on where the AD is heading and, second for the vehicle, in case of a driver misoperation (e.g. sudden hands off the steering wheel). This load can be designed and tuned according to the steering feel, safety and any other objectives.

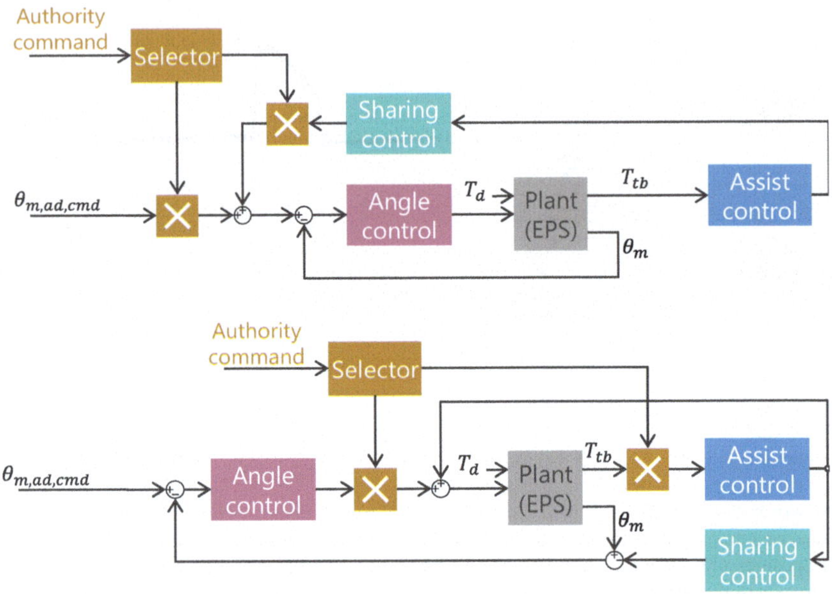

Figure 6: Haptic shared control frameworks: AA configuration (top), TA configuration (bottom). The selector functions enable an external input to allocate the steering mode, for example from the cooperative control.

Extension for cooperative control

The previous section presented two EPS control configurations that enable AD and MD to coexist. There is no switch or weighting function for setting the steering mode. In these configurations, the driver and the automation system can both steer at any time and in any driver or traffic condition. For safety reasons, a mean to allocate the steering control authority is required. This will allow prioritization of manual, automated or shared mode according to a higher level instruction. The tactical and strategical controls are responsible for evaluating the traffic context as well as the driver awareness and readiness necessary in the priority decision process. This is called cooperative control.

As presented, the shared controls lack a mean to select and tune the steering mode (MD, AD or any level of blending). One way to allow input from the cooperative control is to extend the proposed configurations by adding a control authority selector after the AD and MD commands (Figure 6). Continuous and smooth fade in/out functions are preferred to ensure seamless prioritization.

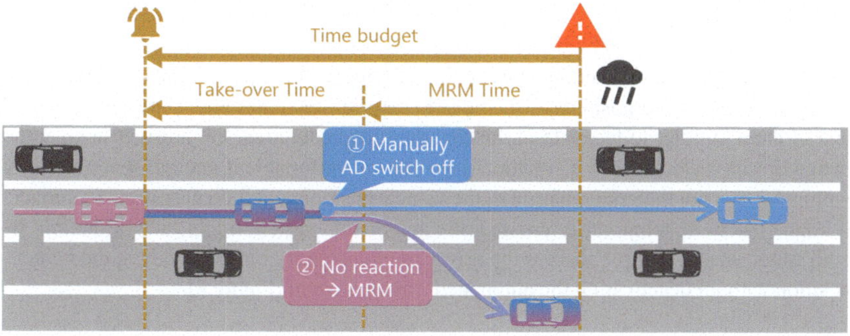

Figure 7: Application to a TOR. Shared control is set immediately after the release of the TOR enabling the driver to steer at any time. Only confirmed operation of the driver can turn-off the AD mode. Furthermore, shared control can remain active when launching the minimum risk manoeuvre in case of no reaction from the driver.

Examples of applications

The TA shared control is intended mainly for application in vehicles equipped with an EPS while AA configuration is dedicated primarily to SBW system. This section presents various examples of applications which benefits from sharing the control of the vehicle lateral dynamics. It shows that well-known steering functions required in highly automated vehicles don't need separate developments. The proposed shared controls enable addressing most of these issues within a single and generic framework.

Take-over request (TOR)

The TOR is a demand, sent by the vehicle to the driver, asking to regain manual control in the case of an anticipated limitation of the automation system (Figure 7). Safe operation of the TOR has generated significant attention due to the complexity in setting a reasonable time frame for the driver to recover awareness of the traffic situation. Shared control offers a partial answer to the problem of authority transfer by avoiding an exclusive interpretation. It can be activated instantly when the TOR is released enabling manual steering without immediate disengagement of the AD mode. When the driver feels confident in regaining the vehicle control, he can turn-off the AD mode. Even in the worst case of the driver failing from manually taking the control of the vehicle, the minimum risk manoeuvre can be launched without particular setting of the EPS control. This illustrates the consistent operation of the proposed framework.

Road narrowing

Drivers often encounter road narrowing for example at the location of roadwork. Intuitively, they adjust the position of the vehicle between the lanes and, sometimes, across one of them in order to feel safe in the changing road condition. For a driver in an automated driving vehicle, such a situation might be uncomfortable if enough space remains within the lane but the safety margin of the vehicle position is getting narrower. Moreover, if a lane has to be temporarily crossed, the automation system might face limitation. In both cases, shared control enables the driver to momentarily manually adjust the vehicle position without deactivating the automated driving system (Figure 8).

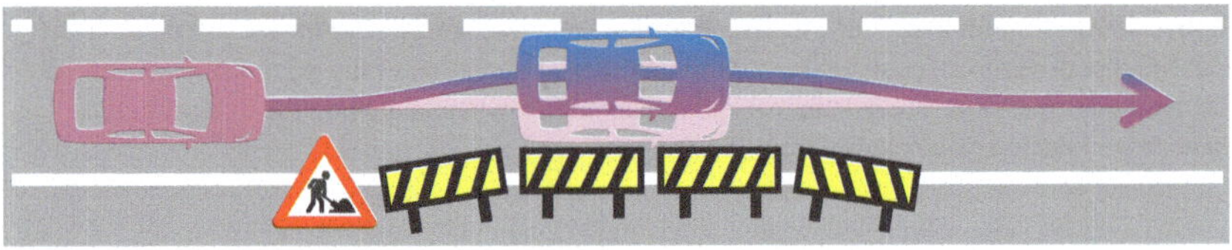

Figure 8: Example of adjustment of vehicle positioning using shared control. During the driver intervention, the AD remains active.

Guidance

Shared control has been presented as offering the possibility for the driver to manually intervene over the automated command. Highly automated vehicles are unlikely to be operated always as such. Indeed, the level of automation will be adjusted depending on the road condition and the automated system capability. Shared control offers also the possibility of developing ADAS (SAE level 1 or 2) for example for continuous or intermittent guidance based on the ability of the vehicle to generate a trajectory to be followed. Such an ADAS would superpose a small guidance torque to the manual assistance.

Evasive Steering Assist and Automated emergency steering (AES)

Evasive steering assist and automated emergency steering are technologies that are gaining particular attention recently because of their potential in providing additional contribution to vehicle and road safety [9]. The former technology assists the driver after detecting his intention in initiating an evasive manoeuvre. A small torque is added to that of the driver for facilitating the collision avoidance. The responsibility remains that of the driver as the function is active only while manually steering. Similarly to the application of guidance, shared control can be applied to superpose the evasive assist torque. AES is more complicated as the driver is not responsible anymore. Technically, such a function has, at least, two additional constraints. The first is to avoid causing another accident during the emergency manoeuvre especially if lane crossing is required. The second is to ensure no harm to the driver when the steering wheel abruptly makes large angular displacement. SBW system and angle overlay device offer hardware solutions to this second issue but to a cost that is not acceptable to all vehicles. Shared control provides a partial but cost-competitive solution because of the compliance gained for a driver input. Indeed, if an arm lies in the steering wheel, the force felt by the driver will be that generated by the spring in the sharing controller (Figure 4) rather than that developed from the maximum torque of the assist motor. Driver injury is prevented at the cost of a deviation from the AD trajectory.

Override function for SBW systems

Driver overriding or taking over the automation system to manually control the vehicle requires a transition between the two exclusive modes of automated and manual driving as described in the first section of this paper. In terms of driver comfort, one of the benefits of SBW system in an automated driving vehicle is the possibility of locking the steering wheel position. Conversely, a drawback of the fixed steering wheel arises during a driver takeover operation while the steering wheel is not aligned to the wheels. If an angular synchronization is required, a robot like control (e.g. position control) is performed at first to cancel the angular offset. Such operation is critical for two contradictory reasons. The time allotted for the angular synchronization should be short enough so as not to constrain the driver in his manoeuvre but long enough to prevent injury caused by too fast motion of the steering wheel. AA shared control provides a solution to these issues. As shown in Figure 9, it can be implemented in a two-stage form with the first stage applied to the rack (gear) and the second to the steering wheel with the reaction force unit (column). Both stages are angle controlled but compliant to the driver input. In this implementation, the angular synchronization is triggered by the detection of the hands on the steering wheel (e.g. using the hands-on detection function described in [10]). A smooth function is used to align the steering wheel (column) to the wheels (gear). The first advantage of applying the proposed control is that a driver intervention is enabled at any time throughout the synchronization. The second advantage comes from the steering wheel compliance that guarantees no harm to the driver during the fast motion. In practice, the following two extreme patterns are possible. First, the driver holds the steering wheel without applying much torque. In this case, the steering wheel aligns to the wheels. Second, the driver holds firmly the steering wheel. Here, the wheels align to the steering wheel.

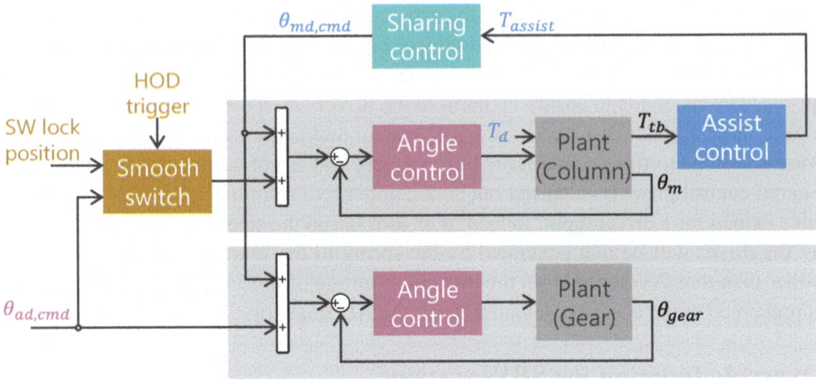

Figure 9: Example of haptic shared control applied to a SBW system. The selector function for the driver input is replaced by a smooth switch function triggered by the detected hands state. For the sake of clarity, the selector functions for the steering authority allocation are not shown.

Conclusion

The proposed haptic shared control has been presented as a generic control architecture for EPS where both manual and automated steering can coexist. Because of its ability to tune the steering control mode (MD, AD or any level of sharing), it is compatible with the higher level of a cooperative control. It provides solutions to various functional requirements of EPS in highly automated vehicles within a single, computational efficient and consistent framework.

Technically, the concept is cost-efficient and has a good adaptability as based on the existing hardware. Mature technologies of steering assistance and angle control can be implemented in their original form and can still be independently improved. Only an additional controller of limited computational weight is required to blend the two controls. The proposed shared control has an input dedicated to allocating the steering authority for consistent implementation with the cooperative control.

From a driver perspective, the haptic shared control enables AD and MD to be active at the same time without the need for manual switching. Naturally, the driver can steer over the AD operation. Conversely, he feels the AD heading direction through haptic feedback.

Bibliography

[1] F. Flemisch, et. al., Shared Control is the Sharp Edge of Cooperation: Toward a Common Framework of Joint Action, Shared Control and Human Machine Cooperation, 13th IFAC Symposium on Analysis, design and Evaluation of Human-Machine Systems, Kyoto, 2016

[2] Surface Vehicle Recommended Practice – Taxonomy and Definitions for Terms Related to Driving Automation Systems for On-Road Motor Vehicles, J3016, SAE International, 2014

[3] A. Seeck, Automated Driving: German Status and Euro NCAP's Approach, NCAP & Car Safety Forum, Tokyo, 2018.

[4] M. Moreillon, et. al., Transfer of Lateral Control Authority for Automated Vehicles, JSAE Congress (Autumn), Nagoya, 2018.

[5] M. A. Benloucif, et. al., A New Scheme for Haptic Shared Lateral Control in Highway Driving Using Trajectory Planning, 20th IFAC World Congress, Toulouse, 2017.

[6] V. Cerone, et. al., Combined Automatic Lane-Keeping and Driver's Steering Through a 2-DOF Control Strategy, IEEE Transactions on Control Systems Technology (Volume: 17, Issue: 1), 2009.

[7] D. A. Abbink, et. al., Haptic Shared Control: Smoothly Shifting Control Authority, Cognition, Technology & Work (Volume: 14, Issue: 1), 2012.

[8] J. Klessing, et. al., Steering on Demand for Dual-Mode Vehicles, chassis.tech plus, Munich, 2018.

[9] Euro 2025 Roadmap, In Pursuit of Vision Zero, Euro NCAP, 2017.

[10] M. Moreillon, et. al., Detection of the Driver Hands On and Off the Steering Wheel for ADAS and Autonomous Driving, 7th International Munich Chassis Symposium, Springer, 2016.

The manufacturer's authorised representative in the EU is Springer Nature Customer Service Centre GmbH, Europaplatz 3, 69115 Heidelberg, Germany. If you have any concerns regarding our products, please contact ProductSafety@springernature.com

Printed and bound by CPI Group (UK) Ltd, Croydon, CR0 4YY
19/08/2026
02200177-0003